Application of Novel Plasmonic Nanomaterials on SERS

Application of Novel Plasmonic Nanomaterials on SERS

Editor

Grégory Barbillon

MDPI • Basel • Beijing • Wuhan • Barcelona • Belgrade • Manchester • Tokyo • Cluj • Tianjin

Editor
Grégory Barbillon
EPF-Ecole d'Ingénieurs
(Faculty of Engineering)
France

Editorial Office
MDPI
St. Alban-Anlage 66
4052 Basel, Switzerland

This is a reprint of articles from the Special Issue published online in the open access journal *Nanomaterials* (ISSN 2079-4991) (available at: https://www.mdpi.com/journal/nanomaterials/special_issues/Plasmonic_SERS).

For citation purposes, cite each article independently as indicated on the article page online and as indicated below:

LastName, A.A.; LastName, B.B.; LastName, C.C. Article Title. *Journal Name* **Year**, *Volume Number*, Page Range.

ISBN 978-3-03943-919-5 (Hbk)
ISBN 978-3-03943-920-1 (PDF)

© 2020 by the authors. Articles in this book are Open Access and distributed under the Creative Commons Attribution (CC BY) license, which allows users to download, copy and build upon published articles, as long as the author and publisher are properly credited, which ensures maximum dissemination and a wider impact of our publications.

The book as a whole is distributed by MDPI under the terms and conditions of the Creative Commons license CC BY-NC-ND.

Contents

About the Editor vii

Grégory Barbillon
Application of Novel Plasmonic Nanomaterials on SERS
Reprinted from: *Nanomaterials* **2020**, *10*, 2308, doi:10.3390/nano10112308 1

Hossein Dizajghorbani-Aghdam, Thomas S. Miller, Rasoul Malekfar and Paul F. McMillan
SERS-Active Cu Nanoparticles on Carbon Nitride Support Fabricated Using Pulsed Laser Ablation
Reprinted from: *Nanomaterials* **2019**, *9*, 1223, doi:10.3390/nano9091223 5

Longkun Yang, Jingran Yang, Yuanyuan Li, Pan Li, Xiaojuan Chen and Zhipeng Li
Controlling the Morphologies of Silver Aggregates by Laser-Induced Synthesis for Optimal SERS Detection
Reprinted from: *Nanomaterials* **2019**, *9*, 1529, doi:10.3390/nano9111529 21

Yu-Chung Chang, Bo-Han Huang and Tsung-Hsien Lin
Surface-Enhanced Raman Scattering and Fluorescence on Gold Nanogratings
Reprinted from: *Nanomaterials* **2020**, *10*, 776, doi:10.3390/nano10040776 31

Iman Ragheb, Macilia Braïk, Stéphanie Lau-Truong, Abderrahmane Belkhir, Anna Rumyantseva, Sergei Kostcheev, Pierre-Michel Adam, Alexandre Chevillot-Biraud, Georges Lévi, Jean Aubard, Leïla Boubekeur-Lecaque and Nordin Félidj
Surface Enhanced Raman Scattering on Regular Arrays of Gold Nanostructures: Impact of Long-Range Interactions and the Surrounding Medium
Reprinted from: *Nanomaterials* **2020**, *10*, 2201, doi:10.3390/nano10112201 45

Grégory Barbillon, Andrey Ivanov and Andrey K. Sarychev
Hybrid Au/Si Disk-Shaped Nanoresonators on Gold Film for Amplified SERS Chemical Sensing
Reprinted from: *Nanomaterials* **2019**, *9*, 1588, doi:10.3390/nano9111588 59

Grégory Barbillon
Latest Novelties on Plasmonic and Non-Plasmonic Nanomaterials for SERS Sensing
Reprinted from: *Nanomaterials* **2020**, *10*, 1200, doi:10.3390/nano10061200 71

About the Editor

Grégory Barbillon completed his PhD in Physics (2007) with greatest distinction at the University of Technology of Troyes (France). He then obtained his Habilitation (HDR) in Physics (2013) at the University of Paris Sud (Orsay, France). He has been a Professor of Physics at the Faculty of Engineering "EPF-Ecole d'Ingénieurs" (Sceaux, France) since his appointment in September 2017. His research interests are focused on plasmonics, nano-optics, non-linear optics, biosensing, optical sensing, condensed matter physics, nanophotonics, nanotechnology, surface-enhanced spectroscopies, sum-frequency generation spectroscopy, materials chemistry, physical chemistry, and fluorescence.

Editorial

Application of Novel Plasmonic Nanomaterials on SERS

Grégory Barbillon

EPF-Ecole d'Ingénieurs, 3 bis rue Lakanal, 92330 Sceaux, France; gregory.barbillon@epf.fr

Received: 11 November 2020; Accepted: 14 November 2020; Published: 22 November 2020

During these past two decades, the fabrication of ultrasensitive surface-enhanced Raman scattering (SERS) substrates has exploded by using novel plasmonic materials such bimetallic materials (e.g., Au/Ag) [1–4], hybrid materials (e.g., metal/semiconductor) [5–8], and also new designs of plasmonic nanostructures (e.g., nanoparticle self-assembly [9–11]). These novel plasmonic nanomaterials can allow a better confinement of the electric field and thus induce an enhancement of the SERS signal (electromagnetic contribution [12,13]) by adjusting, for instance, the size, shape, periodicity, nanoparticle self-assembly, and nanomaterials' nature. These nanomaterials can also enhance the charge transfer (electrons; chemical contribution) to increase the SERS signal [14,15]. Furthermore, other materials are appeared for SERS applications such as metal oxides [16,17]. Other directions for the SERS field also emerged such as the SERS effect induced by high pressure [18,19], and the photo-induced enhanced Raman spectroscopy [20–22]. Thus, this special issue is dedicated to introducing recent advances and insights in these novel plasmonic nanomaterials applied to the fabrication of highly sensitive SERS substrates for chemical and biological sensing.

This special issue is formed of 5 research articles, and 1 review article. The first part of this latter is devoted to the novel methods of fabrication of plasmonic nanoparticles or nanostructures for SERS [23–25]. Dizajghrobani-Aghdam et al. demonstrated an alternative method of fabrication of metallic nanoparticles by employing pulsed laser ablation, and these hybrid plasmonic nanostructures have presented significant enhancements of the Raman signal [23]. Furthermore, Yang et al. presented the direct fabrication of SERS substrates by using an *in situ* photochemical method of reduction. High enhancements of the Raman signal were obtained with these SERS substrates [24]. To finish this first part, Chang et al. proposed a simpler method of electron beam lithography in order to realize SERS substrates by removing the photoresist lift-off step [25]. In the second part, the presented domain is devoted to the impact of long-range interactions and the surrounding medium on the SERS effect investigated by Ragheb et al. [26]. In the last part, the addressed domains are dedicated to novel plasmonic and non-plasmonic nanomaterials for SERS sensing [27,28]. Barbillon et al. demonstrated the enhancement of the Raman signal with hybrid nanostructures on a metallic film [27]. To finish this part and this special issue on novel plasmonic nanomaterials applied to the SERS field, Barbillon presented a short review on plasmonic and non-plasmonic nanomaterials for SERS sensing [28].

For performing the special issue entitled "Application of Novel Plasmonic Nanomaterials on SERS", a couple of contributions has been obtained from excellent-quality authors originate from worldwide. I would like to acknowledge all these authors as well as the whole editorial office of the journal "Nanomaterials" for their great support and help in the management process of the article submissions and other associated tasks. To finish, I hope that you will find interesting this special issue devoted to novel plasmonic nanomaterials for the SERS field, which is targeted to the students or researchers who are or wish to imply in this field.

Funding: This research received no external funding.

Conflicts of Interest: The author declares no conflict of interest.

References

1. Song, J.B.; Duan, B.; Wang, C.X.; Zhou, J.J.; Pu, L.; Fang, Z.; Wang, P.; Lim, T.T.; Duan, D.W. SERS-Encoded Nanogapped Plasmonic Nanoparticles: Growth of Metallic Nanoshell by Templating Redox-Active Polymer Brushes. *J. Am. Chem. Soc.* **2014**, *136*, 6838–6841. [CrossRef]
2. Yang, Y.; Liu, J.; Fu, Z.W.; Qin, D. Galvanic replacement-free deposition of Au on Ag for core-shell nanocubes with enhanced chemical stability and SERS activity. *J. Am. Chem. Soc.* **2014**, *136*, 8153–8156. [CrossRef]
3. Feng, J.J.; Wu, X.L.; Ma, W.; Kuang, H.; Xu, L.G.; Xu, C.L. A SERS active bimetallic core-satellite nanostructure for the ultrasensitive detection of Mucin-1. *Chem. Commun.* **2015**, *51*, 14761–14763. [CrossRef]
4. Zhang, Y.; Yang, P.; Habeeb Muhammed, M.A.; Alsaiari, S.K.; Moosa, B.; Almalik, A.; Kumar, A.; Ringe, E.; Kashab, N.M. Tunable and Linker Free Nanogaps in Core-Shell Plasmonic Nanorods for Selective and Quantitative Detection of Circulating Tumor Cells by SERS. *ACS Appl. Mater. Interfaces* **2017**, *9*, 37597–37605. [CrossRef]
5. Bryche, J.-F.; Bélier, B.; Bartenlian, B.; Barbillon, G. Low-cost SERS substrates composed of hybrid nanoskittles for a highly sensitive sensing of chemical molecules. *Sens. Actuators B* **2017**, *239*, 795–799. [CrossRef]
6. Wu, J.; Du, Y.; Wang, C.; Bai, S.; Zhang, T.; Chen, T.; Hu, A. Reusable and long-life 3D Ag nanoparticles coated Si nanowire array as sensitive SERS substrate. *Appl. Surf. Sci.* **2019**, *494*, 583–590. [CrossRef]
7. Yang, M.S.; Yu, J.; Lei, F.C.; Zhou, H.; Wei, Y.S.; Man, B.Y.; Zhang, C.; Li, C.H.; Ren, J.F.; Yuan, X.B. Synthesis of low-cost 3D-porous ZnO/Ag SERS-active substrate with ultrasensitive and repeatable detectability. *Sens. Actuators B* **2018**, *256*, 268–275. [CrossRef]
8. Graniel, O.; Iatsunskyi, I.; Coy, E.; Humbert, C.; Barbillon, G.; Michel, T.; Maurin, D.; Balme, S.; Miele, P.; Bechelany, M. Au-covered hollow urchin-like ZnO nanostructures for surface-enhanced Raman scattering sensing. *J. Mater. Chem. C* **2019**, *7*, 15066–15073. [CrossRef]
9. Matricardi, C.; Hanske, C.; Garcia-Pomar, J.L.; Langer, J.; Mihi, A.; Liz-Marzan, L.M. Gold Nanoparticle Plasmonic Superlattices as Surface-Enhanced Raman Spectroscopy Substrates. *ACS Nano* **2018**, *12*, 8531–8539. [CrossRef] [PubMed]
10. Volk, K.; Fitzgerald, J.P.S.; Ruckdeschel, P.; Retsch, M.; König, T.A.F.; Karg, M. Reversible Tuning of Visible Wavelength Surface Lattice Resonances in Self-Assembled Hybrid Monolayers. *Adv. Opt. Mater.* **2017**, *5*, 1600971. [CrossRef]
11. Greybush, N.J.; Liberal, I.; Malassis, L.; Kikkawa, J.M.; Engheta, N.; Murray, C.B.; Kagan, C.R. Plasmon Resonances in Self-Assembled Two-Dimensional Au Nanocrystal Metamolecules. *ACS Nano* **2017**, *11*, 2917–2927. [CrossRef] [PubMed]
12. Itoh, T.; Yamamoto, Y.S.; Ozaki, Y. Plasmon-enhanced spectroscopy of absorption and spontaneous emissions explained using cavity quantum optics. *Chem. Soc. Rev.* **2017**, *46*, 3904–3921. [CrossRef] [PubMed]
13. Ding, S.-Y.; You, E.-M.; Tian, Z.-Q.; Moskovits, M. Electromagnetic theories of surface-enhanced Raman spectroscopy. *Chem. Soc. Rev.* **2017**, *46*, 4042–4076. [CrossRef] [PubMed]
14. Jensen, L.; Aikens, C.M.; Schatz, G.C. Electronic structure methods for studying surface-enhanced Raman scattering. *Chem. Soc. Rev.* **2008**, *37*, 1061–1073. [CrossRef]
15. Alessandri, I.; Lombardi, J.R. Enhanced Raman Scattering with Dielectrics. *Chem. Rev.* **2016**, *116*, 14921–14981. [CrossRef]
16. Cong, S.; Yuan, Y.; Chen, Z.; Hou, J.; Yang, M.; Su, Y.; Zhang, Y.; Li, L.; Li, Q.; Geng, F.; et al. Noble metal-comparable SERS enhancement from semiconducting metal oxides by making oxygen vacancies. *Nat. Commun.* **2015**, *6*, 7800. [CrossRef]
17. Liu, W.; Bai, H.; Li, X.; Li, W.; Zhai, J.; Li, J.; Xi, G. Improved Surface-Enhanced Raman Spectroscopy Sensitivity on Metallic Tungsten Oxide by the Synergistic Effect of Surface Plasmon Resonance Coupling and Charge Transfer. *J. Phys. Chem. Lett.* **2018**, *9*, 4096–4100. [CrossRef]
18. Sun, H.H.; Yao, M.G.; Song, Y.P.; Zhu, L.Y.; Dong, J.J.; Liu, R.; Li, P.; Zhao, B.; Liu, B.B. Pressure-induced SERS enhancement in a MoS_2/Au/R6G system by a two-step charge transfer process. *Nanoscale* **2019**, *11*, 21493–21501. [CrossRef]
19. Barbillon, G. Nanoplasmonics in High Pressure Environment. *Photonics* **2020**, *7*, 53. [CrossRef]

20. Ben-Jaber, S.; Peveler, W.J.; Quesada-Cabrera, R.; Cortés, E.; Sotelo-Vazquez, C.; Abdul-Karim, N.; Maier, S.A.; Parkin, I.P. Photo-induced enhanced Raman spectroscopy for universal ultra-trace detection of explosives, pollutants and biomolecules. *Nat. Commun.* **2016**, *7*, 12189. [CrossRef]
21. Glass, D.; Cortés, E.; Ben-Jaber, S.; Brick, T.; Peveler, W.J.; Blackman, C.S.; Howle, C.R.; Quesada-Cabrera, R.; Parkin, I.P.; Maier, S.A. Dynamics of Photo-Induced Surface Oxygen Vacancies in Metal-Oxide Semiconductors Studied Under Ambient Conditions. *Adv. Sci.* **2019**, *6*, 1901841. [CrossRef] [PubMed]
22. Barbillon, G.; Noblet, T.; Humbert, C. Highly crystalline ZnO film decorated with gold nanospheres for PIERS chemical sensing. *Phys. Chem. Chem. Phys.* **2020**, *22*, 21000–21004. [CrossRef] [PubMed]
23. Dizajghorbani-Aghdam, H.; Miller, T.S.; Malekfar, R.; McMillan, P.F. SERS-Active Cu Nanoparticles on Carbon Nitride Support Fabricated Using Pulsed Laser Ablation. *Nanomaterials* **2019**, *9*, 1223. [CrossRef] [PubMed]
24. Yang, L.; Yang, J.; Li, Y.; Li, P.; Chen, X.; Li, Z. Controlling the Morphologies of Silver Aggregates by Laser-Induced Synthesis for Optimal SERS Detection. *Nanomaterials* **2019**, *9*, 1529. [CrossRef]
25. Chang, Y.C.; Huang, B.-H.; Lin, T.-H. Surface-Enhanced Raman Scattering and Fluorescence on Gold Nanogratings. *Nanomaterials* **2020**, *10*, 776. [CrossRef]
26. Ragheb, I.; Braïk, M.; Lau-Trong, S.; Belkir, A.; Rumyantseva, A.; Kostcheev, S; Adam, P.-M.; Chevillot-Biraud, A.; Lévi, G.; Aubard, J.; et al. Surface Enhanced Raman Scattering on Regular Arrays of Gold Nanostructures: Impact of Long-Range Interactions and the Surrounding Medium. *Nanomaterials* **2020**, *10*, 2201. [CrossRef]
27. Barbillon, G.; Ivanov, A.; Sarychev, A.K. Hybrid Au/Si Disk-Shaped Nanoresonators on Gold Film for Amplified SERS Chemical Sensing. *Nanomaterials* **2019**, *9*, 1588. [CrossRef]
28. Barbillon, G. Latest Novelties on Plasmonic and Non-Plasmonic Nanomaterials for SERS sensing. *Nanomaterials* **2020**, *10*, 1200. [CrossRef]

Publisher's Note: MDPI stays neutral with regard to jurisdictional claims in published maps and institutional affiliations.

© 2020 by the authors. Licensee MDPI, Basel, Switzerland. This article is an open access article distributed under the terms and conditions of the Creative Commons Attribution (CC BY) license (http://creativecommons.org/licenses/by/4.0/).

Article

SERS-Active Cu Nanoparticles on Carbon Nitride Support Fabricated Using Pulsed Laser Ablation

Hossein Dizajghorbani-Aghdam [1], Thomas S. Miller [2], Rasoul Malekfar [1,*] and Paul F. McMillan [3,*]

1. Atomic and Molecular Group, Physics Department, Faculty of Basic Sciences, Tarbiat Modares University, Tehran 14115-175, Iran
2. Electrochemical Innovation Lab, Department of Chemical Engineering, University College London, Torrington Place, London WC1E 7JE, UK
3. Department of Chemistry, Christopher Ingold Laboratories, University College London, 20 Gordon Street, London WC1H 0AJ, UK
* Correspondence: malekfar@modares.ac.ir (R.M.); p.f.mcmillan@ucl.ac.uk (P.F.M)

Received: 17 July 2019; Accepted: 23 August 2019; Published: 29 August 2019

Abstract: We report a single-step route to co-deposit Cu nanoparticles with a graphitic carbon nitride (gCN) support using nanosecond Ce:Nd:YAG pulsed laser ablation from a Cu metal target coated using acetonitrile (CH_3CN). The resulting Cu/gCN hybrids showed strong optical absorption in the visible to near-IR range and exhibited surface-enhanced Raman or resonance Raman scattering (SERS or SERRS) enhancement for crystal violet (CV), methylene blue (MB), and rhodamine 6G (R6G) used as probe analyte molecules adsorbed on the surface. We have characterized the Cu nanoparticles and the nature of the gCN support materials using a range of spectroscopic, structural, and compositional analysis techniques.

Keywords: pulsed laser ablation; acetonitrile (CH_3CN); Cu/gCN hybrids; localized surface plasmon resonance (LSPR); surface enhanced Raman scattering (SERS); surface enhanced resonance Raman scattering (SERRS)

1. Introduction

Surface-enhanced Raman spectroscopy (SERS) is a highly sensitive spectroscopic technique used to detect and study vibrational modes of molecules adsorbed on metal or semiconducting surfaces or nanoparticles (NP) [1–7]. SERS intensity enhancement factors (EFs) by up to 14 orders of magnitude arise from a combination of electromagnetic (EM) and chemical (CM) mechanisms [8,9]. EM enhancement factors ($\sim 10^7$–10^8) are dominant for metallic materials [10,11] compared with the CM enhancement ($\sim 10^3$) operating mainly among semiconductors [12–15]. Typical SERS active supports include noble metals (Au, Ag, Cu) that develop localized surface plasmon resonances (SPR) in the visible range, giving rise to the photon-vibrational coupling effects leading to SERS with EM enhancement [11,16]. In addition, if the adsorbed analyte contains a chromophore with a UV-visible absorption close to the SPR wavelength, additional enhancement can be observed due to resonance Raman (SERRS) effects [17].

Ag and Au NPs have been mainly investigated as the most highly active SERS substrates [11,18–21]. Cu NPs exhibit similar SPR properties but can show lower chemical stability toward adsorbate species [16,22–25]. Cu NPs are readily produced using solution chemistry techniques, but these approaches can lead to surface contamination due to atmospheric oxidation, as well as from the addition of surfactants such as polyethylene glycol, sodium dodecyl sulfate, etc. [26–28], thus hindering their use for SERS, as well as for biomedical and catalytic applications [29,30]. Laser ablation from a pure metal target in the presence of a liquid medium has been shown to provide a useful alternative to obtain SERS-active NPs [24,25,31]. Using liquids such as H_2O or acetone results in the formation of Cu oxides [32–34] that reduce the SERS activity and affect the chemical stability [35]. Here, we explored the use of the oxygen-free liquids toluene (C_7H_8) and acetonitrile (CH_3CN) for the Cu laser ablation experiments. The aromatic hydrocarbon toluene produced Cu NPs surrounded by an amorphous carbon matrix that resulted in quenching of the SPR. Similar results have been reported for Au NPs [36,37]. These effects could be reduced by incorporating N atoms in the ablation medium and the resulting solid matrix surrounding and supporting the Cu NPs. Previous studies have described the laser ablation of Cu and other transition metal elements using acetonitrile to obtain metal NPs deposited along with a carbon nitride support, but the nature of the support material obtained has not yet been fully established [38–42]. Here, we characterized the Cu NPs and the carbon nitride matrix, that we established to have a polymeric to graphitic character based on mainly sp^2-bonded C and N species (labelled gCN for convenience), using a range of spectroscopic, diffraction, and imaging techniques. We evaluated the performance of the Cu/gCN hybrid nanomaterials for their surface-enhanced Raman or resonance Raman scattering (SERS or SERRS) performance using crystal violet (CV), methylene blue (MB), and rhodamine 6G (R6G) as probe analyte molecules adsorbed on the NP surface.

Different classes of carbon nitride materials are currently being investigated for applications including catalysis, photocatalysis, electrochemical energy storage, and conversion, and as supports for catalytically active metal NPs [43–46]. Most work to date is focused on semiconducting compounds with $C_xN_yH_z$ compositions and polymeric to layered ("graphitic") structures are being developed. Although these materials are often cited as graphitic (g-) C_3N_4, this particular stoichiometry and layered structure have only been characterized in a single study that described a triazine-based graphitic carbon nitride (TGCN), produced using the polymerization of dicyandiamide (DCDA: $C_2N_4H_4$) in a LiCl:KCl molten salt medium [47]. That process normally results in crystalline polytriazine imide (PTI) solids containing intercalated Li^+ and Cl^- ions [48]; the TGCN material consisted of dark-colored flakes formed at the molten surface or deposited on the walls of the reaction vessel. However, related studies have found that the layered gCN materials deposited from above the DCDA/molten salt mixture under analogous reaction conditions had significantly greater C:N ratios, close to 1.3 [49,50]. That composition is similar to the gCN supports formed in our laser ablation study, described below. The wide range of materials described as carbon nitrides include N-doped graphite and graphene produced by chemical or physical vapor deposition exhibit semi-metallic to semiconducting behavior, with optoelectronic properties determined by the relative C:N ratio and their distribution within the sp^2-bonded structures [51–53]. Such materials have been formed via the laser ablation of graphite targets in N_2 or NH_3 atmospheres, or by the laser-assisted reactions of precursors in gas and liquid media, giving rise to materials containing up to ≈20 at% nitrogen in the resulting amorphous to nanocrystalline films [54–60]. There have been reports of embedding Cu NPs in carbon nitride materials for functional applications. Hydrothermal processing in Cu-containing ionic liquids produced a Cu/gCN electrocatalyst that showed catalytic activity for the oxygen reduction reaction (ORR) critical for fuel cell operations [61], while reacting $CuCl_2$ with melamine in methanol formed supramolecular networks and carbon nitride nanosheets and nanorods with coordinated Cu^{2+} ions resulting in enhanced catalytic and visible light photocatalysis properties [62–66].

These last investigations indicate that electronic interactions between Cu NPs and supporting carbon nitride matrices can lead to the cooperative enhancement of catalytic and photocatalytic effects. Such interactions could be harnessed to develop other optical properties, including SERS enhancement of the vibrational signatures of molecular species adsorbed on the surface. Studies have reported combining semiconductor NPs (such as TiO_2) with plasmonically active NPs to stabilize and enhance the performance of SERS substrates [67–69]. Jiang et al. described a hybrid Ag/gCN material with a stable high-performance SERS activity, demonstrating strong optoelectronic interactions that promote the SERS enhancement along with charge transfer effects that protect the Ag NPs from oxidation [70,71]. Here we applied pulsed laser ablation of a Cu target coated with acetonitrile (CH_3CN) to create Cu/gCN hybrid nanomaterials that were characterized using a range of spectroscopic, diffraction, and imaging techniques, and we studied their SERS/SERRS activity for the probe analyte molecules crystal violet (CV), methylene blue (MB), and rhodamine 6G (R6G).

2. Materials and Methods

A high-purity copper plate (Cu 99.98%) with a 2-mm thickness coated with acetonitrile served as the target for the laser ablation experiments. High-purity, methylene blue (MB), crystal violet (CV) (99.5%), and rhodamine 6G (R6G) (≥95%) were purchased from Sigma Aldrich (Darmstadt, Germany) and used as SERS probe analytes. Bulk elemental analysis was carried out using an Elementar-Vario EL III system (Hanau, Germany) and X-ray photoelectron spectroscopy (XPS) was carried out using a Thermo Scientific K-Alpha instrument (Waltham Massachusetts, U.S.A) with a monochromated Al Kα source (1486.6 eV). Transmission electron microscopy (TEM) was carried out using Zeiss-EM10C-80 kV (Germany) and JEOL JEM2010-200 kV (Tokyo, Japan) instruments. The micrographs were analyzed using ImageJ software (Version 6, California, U.S.A) [72] and by manual counting to determine the mean size of NPs in the TEM images. The average Cu NP sizes were determined to be ≈6 nm, slightly lower than those estimated from Scherrer broadening of the X-ray powder diffraction data. X-ray diffraction (XRD) patterns were recorded using an X-Pert MPD (Philips, Eindhoven, Netherlands) X-ray diffractometer with Co Kα-radiation (λ = 1.79026 Å). Ultraviolet–visible (UV/vis) absorption spectra were obtained using a PG T80+ spectrophotometer (PG Instruments, Coventry, UK), Fourier Transform infrared (FTIR) spectra with a Thermo Nicolet NEXUS 670 FTIR (Waltham Massachusetts, U.S.A), and Raman spectroscopy including SERS or SERRS effects were recorded using a Thermo Nicolet Almega spectrometer (Waltham Massachusetts, U.S.A) with a 532-nm excitation provided by a Nd:YLF laser.

Pulsed laser deposition was carried out using a Ce:Nd:YAG laser (λ = 1064 nm) with a 10-ns pulse width and a 10-Hz repetition rate. The beam was focused on the Cu plate target using a 10-cm working distance lens. The target was fixed to the bottom of a glass cell and mounted on a rotator in order to avoid deep ablation. The laser pulse energy was held constant at 100 mJ/pulse. The typical diameter of the laser spot at the target was 2 mm and the acetonitrile level on the target surface was 2 mm. The pulsed laser was directed on the rotating target surface for 10 min to obtain a colloidal suspension of ablated nanomaterial particles. The suspension was then dried at 25 °C in air to provide solid powdered materials for further investigation.

To prepare SERS substrates, 1 mg of the synthesized support materials (Cu/gCN) were suspended in 5 mL of the different analyte solutions (CV, MB, and R6G) prepared with concentrations ranging between 10^{-3}–10^{-8} mol/L in acetonitrile and stirred for 3 h to allow binding between the analyte and substrate. The analyte-decorated Cu/gCN solutions were centrifuged and washed multiple times with absolute acetonitrile to remove any unabsorbed analyte molecules and then the resulting powders were dispersed into 1 mL acetonitrile. SERS substrates were prepared by dropping aliquots of analyte-decorated Cu/gCN solutions (10 µL) onto a glass slide and the volatile liquid phase was left to evaporate at 25 °C in air to form a uniform coating. All Raman and SERS/SERRS experiments were carried out with the incident laser power maintained at 10 mW for a 60 µm (spot diameter) laser beam focused at different points on the sample surface maintaining a constant integration time of 100 s at a spectral resolution of 4 cm^{-1}. All spectra were measured using 532-nm excitation delivered by the second harmonic of a Nd:YLF laser and were collected in the range of 400–4000 cm^{-1}. All data processing and statistical analysis were carried out using the OMNIC™ software (Version 6) supplied by Thermo Nicolet (Waltham Massachusetts, U.S.A) and intensity values of the spectra were normalized according to the integration time within each data set.

3. Results and Discussion

3.1. Formation and Characterization of Cu/g-CN Materials Produced by Pulsed Laser Ablation in Acetonitrile

The interaction of high-power laser beams with a Cu target can result in several phenomena including heating, melting, and even vaporization of the surface, leading to the formation of plasma. A complex process involving adiabatic plasma expansion combined with shockwaves and the formation of cavitation bubbles result in ablation of Cu NPs that become suspended within the overlying liquid medium. The high temperatures generated by laser beam absorption using the Cu target result in bond-breaking and polymerization reactions occurring within the acetonitrile liquid phase surrounding the ablated metal NPs. Also, radiation from the laser-ablated Cu vapor plume leads to UV absorption and photolysis reactions within the acetonitrile liquid. The combination of these thermal and photolytic reaction effects, accompanied by new C–C/C–N bond formation, results in the formation of a polymerized gCN matrix that then surrounds and supports the Cu NPs formed during laser ablation treatment [38].

We first characterized the chemical composition of the Cu/gCN hybrid materials using bulk elemental (CNH) analysis that revealed a C:N ratio of 1.3, which was larger than those expected for g-C_3N_4 or polymeric to graphitic gCN materials prepared via the thermolysis of organic precursors [43,44]. However, it only slightly exceeded the C:N ratio found for layered gCN materials produced using vapor deposition above a molten salt bath from precursors such as DCDA [49,50]. Examination of the survey C1s and N1s spectra obtained from XPS measurements confirms this observation (Figure 1a–c). The C1s spectra could be fit using three mixed Gaussian–Lorentzian (GL) peaks at 284.3, 286.0, and 288.2 eV. The peak at 288.2 eV appears at a position that is characteristic of sp^2 C atoms bonded within a cyclic triazine or heptazine systems [44], although this constituted only a minor (5.5%) component of the overall C contribution. The peak at 286.0 eV (18.55% of the C contribution) was assigned to C–N bonded species, although it might also contain a contribution from C–O bonded species that are often recorded for gCN materials due to surface oxidation [44,50]. It is probable that the remainder of the C–N bonding was represented by a significant, but unresolvable, contribution to the peak at 284.3 eV, along with a component from adventitious carbon [50,53]. The presence of the C1s peak at 286.0 eV, as well as a significant proportion of that observed at 284.3 eV, confirmed the presence of significant C–N interactions within the sample. The N1s spectrum (Figure 1c) was best fit with two GL peaks, one at 398.4 eV, and a second contribution at 399.5 eV. The smaller peak resembles that observed for melamine and is usually interpreted to be indicative of C–N–H uncondensed amino (–NH_2) groups [44]. We note that the IR spectra discussed below indicate the presence of N–H bonded species. The peak at the higher binding energy (BE) resulted from the remaining C–N species, which included C–N=C bonded units indicated by the peak in the C1s spectrum. The IR and Raman data also indicated the presence of terminal nitrile (–C≡N) units that contributed to the N1s signal. The C1s:N1s peak ratio found from the XPS data was 4.76. This was significantly larger than the value determined using bulk elemental analysis (1.33), but it was certainly affected by a significant contribution from adventitious carbon. However, the fact that the C:N ratio was substantially greater than 1 clearly indicates that both C–C and C–N species were present within the gCN materials. The Cu $2p_{3/2}$ peak (Figure 1d) exhibits a single sharp feature that could be fitted using a GL peak at 933.3 eV. The shape and position of this feature, as well as the absence of higher BE satellites, rule out the presence of Cu(II), but Cu(I) and Cu(0) species both have similar XPS signatures, and it is possible that either or both of these species might be present within the sample [73].

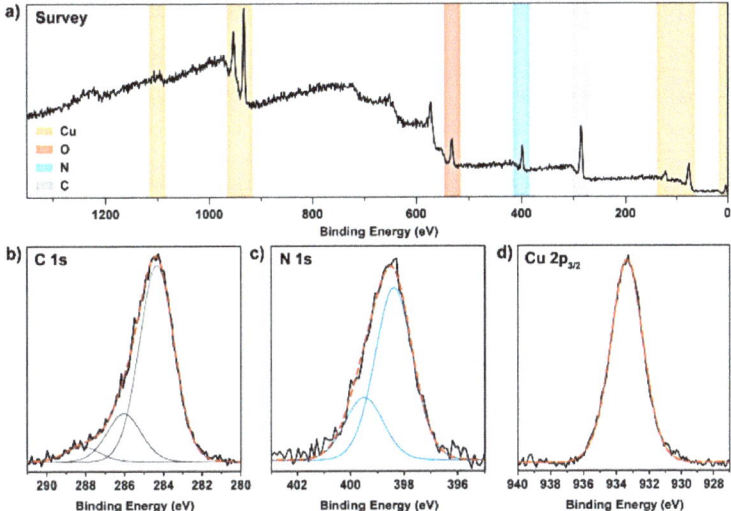

Figure 1. (a) XPS survey spectrum, (b) C1s, (c) N1s, and (d) Cu $2p_{3/2}$ XPS spectra of the Cu/gCN material produced by laser ablation under acetonitrile.

UV-visible absorption spectra of the Cu/gCN nanomaterials obtained in suspension in acetonitrile following the laser ablation synthesis exhibited a peak at ≈580 nm due to the localized surface plasmon resonance (SPR) band of the Cu NPs (Figure 2) [16,24,74].

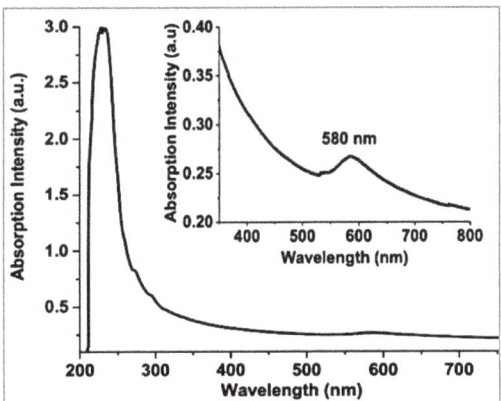

Figure 2. UV–visible absorption spectrum of a typical Cu/gCN nanomaterial synthesized using pulsed laser ablation in acetonitrile. The inset shows a vertically expanded region of the spectrum around the position of the surface plasmon resonance for Cu NPs.

The XRD pattern of dried powders obtained using Co Kα radiation showed weak sharp peaks that were distinct from the broad and more intense features from the amorphous gCN matrix (Figure 3). The weak peaks occurring at 50.7° and 59.2° 2θ (λ = 1.79026 Å) were identified with the {111} and {200} reflections of crystalline Cu (2.09 and 1.08 Å, respectively). A very weak feature could just be distinguished at 89.1° 2θ (1.27 Å) that corresponded to the {220} reflection. Examining the widths of these features indicated an NP size with an average mean diameter ≈32 nm estimated using Scherrer's formula (Table 1). The pattern was dominated by a strong, broad peak at 34.3° 2θ (3.03 Å), along with a weaker feature at 28.9° 2θ (3.59 Å). These resembled the broad reflections observed for polymeric to graphitic carbon nitride materials prepared using thermolysis of organic precursors, where they are commonly interpreted in relation to the {002} and {100} planes of polytriazine to polyheptazine graphitic layered structures, even though the actual materials are known to be incompletely polymerized and contain ribbon-like units along with partially formed sheets [44,75–77]. Similar features were also found for nanocrystalline layered materials with higher C:N ratios close to those observed for the gCN supports created in this study [44,49,50]. We note, however, that the main XRD peak occurred at a value that is ≈11% smaller than the interplanar d_{002} spacing of crystalline graphite (d_{002} = 3.28 Å), or gCN compounds with both low (~3.26 Å) and high C:N ratios (~3.24 Å) [44,49,50,75–77].

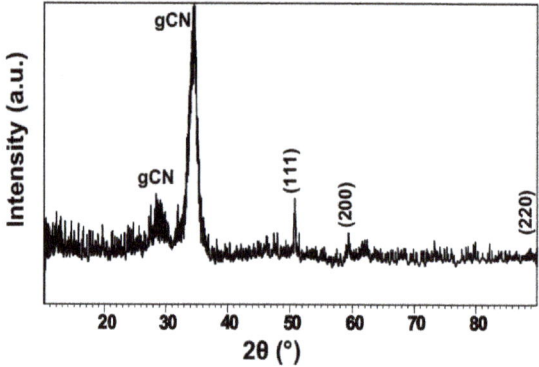

Figure 3. Powder XRD pattern (obtained using Co Kα radiation, λ = 1.79026 Å) of the Cu/gCN material formed using pulsed laser ablation in acetonitrile. The weak sharp peaks due to crystalline Cu NPs are labeled with their {hkl} values. The strong broad features are due to the gCN support generated during the laser ablation synthesis from acetonitrile.

Table 1. Summary of X-ray diffraction data obtained using Co Kα radiation (λ = 1.79026 Å) for a typical Cu/gCN material prepared using pulsed laser ablation in acetonitrile. * The assignment of peaks for gCN is discussed in the text.

Angle (2θ)	{hkl}	FWHM (2θ)	d-spacing (Å)	Height (counts)	Identification
28.9	*	3.3	3.59	27.2	gCN
34.3	*	1.6	3.03	153.9	gCN
50.7	{111}	0.3	2.09	36.9	Cu
59.2	{200}	0.4	1.08	14.6	Cu
89.1	{220}	3.2	1.27	5.4	Cu

Cu NPs were shown to be distributed throughout the gCN matrix via TEM imaging (Figure 4a–d). By counting ≈40 particles, the mean diameter was found to be ≈6 nm by analyzing the images using ImageJ software [72]. A few particles in the field of view were observed to have diameters in the 10–25 nm range. The greater average particle size (≈30 nm) indicated using the XRD results could have been the result of the presence of large NPs within the bulk materials that were not sampled in the TEM experiments, although Scherrer analysis of the X-ray peaks is not always reliable. The gCN supporting matrix revealed a turbostratic texture with an "onion-like" structure developed around the Cu NPs (Figure 4c).

FTIR spectra of a pressed KBr disc of the powdered Cu/gCN sample contained a strong band at 3340 cm^{-1} due to N–H stretching from the gCN material, combined with a contribution from O–H stretching from surface-adsorbed H$_2$O (Figure 5a) [78,79]. The peak at 1640 cm^{-1} was due to molecular H$_2$O bending vibrations [78]. The features at 2850–2950 cm^{-1} indicated the presence of sp^3-hybridized C–H bonds [80]. Features in the 1000–1450 cm^{-1} region can be assigned to C–N and possibly C–O stretching vibrations [44,50]. The sharp peak at 2120 cm^{-1} was due to terminal C≡N bond stretching within the incompletely condensed carbon nitride matrix [79,80]. The strong Raman peak at 2155 cm^{-1} confirmed this assignment (Figure 5b), although it is not clear why there should be a 35 cm^{-1} difference between the IR and Raman data. This could indicate a vibrational interaction between adjacent nitrile groups within the structure giving rise to symmetric and antisymmetric combinations; however, typically the IR mode should occur at a higher wavenumber. Our observation of this feature clearly indicates that the gCN material produced by laser ablation from acetonitrile was not fully polymerized into graphitic sheets. However, the sharp peak in the IR spectrum near 810 cm^{-1} was typical of breathing modes of ring units in carbonaceous and C_xN_y materials. Also, the Raman spectrum contained two main bands at 1346 and 1540 cm^{-1} that resembled the "G" and "D" bands of disordered graphite, amorphous carbon, and graphitic carbon nitride materials, associated with C–N/C–C stretching vibrations within the gCN matrix [44,50,79,80].

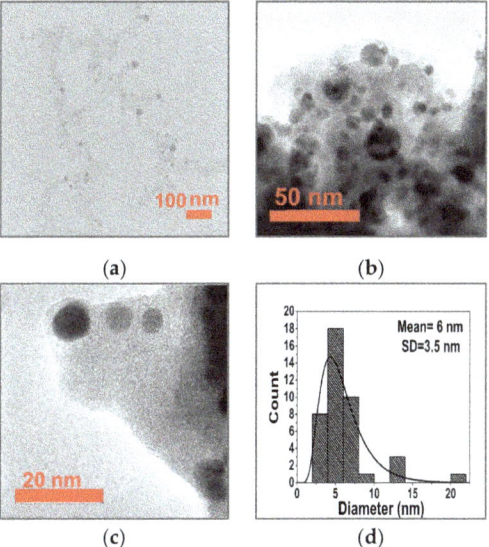

Figure 4. TEM images of Cu NPs embedded in the amorphous gCN support. (**a**) Low-resolution image obtained using a Zeiss-EM10C-80 kV instrument; (**b**,**c**) High-resolution images obtained using a JEOL JEM2010-200 kV TEM. (**d**) Mean size (nm) and distribution by analysis of ≈40 Cu NPs.

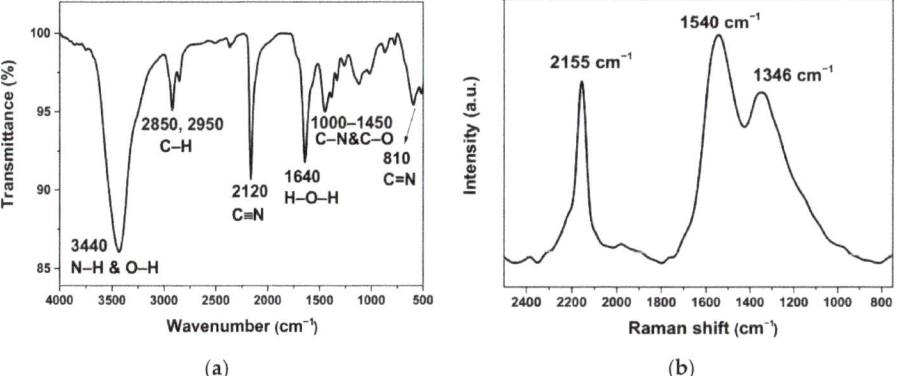

Figure 5. (**a**) FTIR and (**b**) Raman spectra of Cu/gCN hybrid material synthesized using pulsed laser ablation in acetonitrile.

3.2. SERS/SERRS Activity of Analytes on Cu/gCN Hybrid Support Produced Using Pulsed Laser Deposition

We tested the SERS/SERRS activity of the CuNP/g-CN hybrid using the dye molecules CV, MB, and R6G as analyte probes. A drop of each the analyte-decorated Cu/gCN solutions prepared with various due concentrations (10^{-3}–10^{-7} M) was placed on the prepared glass slide substrates separately and allowed to dry before recording surface-enhanced spectra. "Normal" and SERS/SERRS spectra are compared in Figure 6, after subtraction of the background fluorescence (Figure S1). No additional bands were observed from the gCN support material (Figure S2). All spectra were obtained using a total integration time of 100 s from 50 accumulated scans with a 2-s exposure. The spectra shown in Figure 6 have been displaced vertically to more clearly show the SERS/SERRS enhancement.

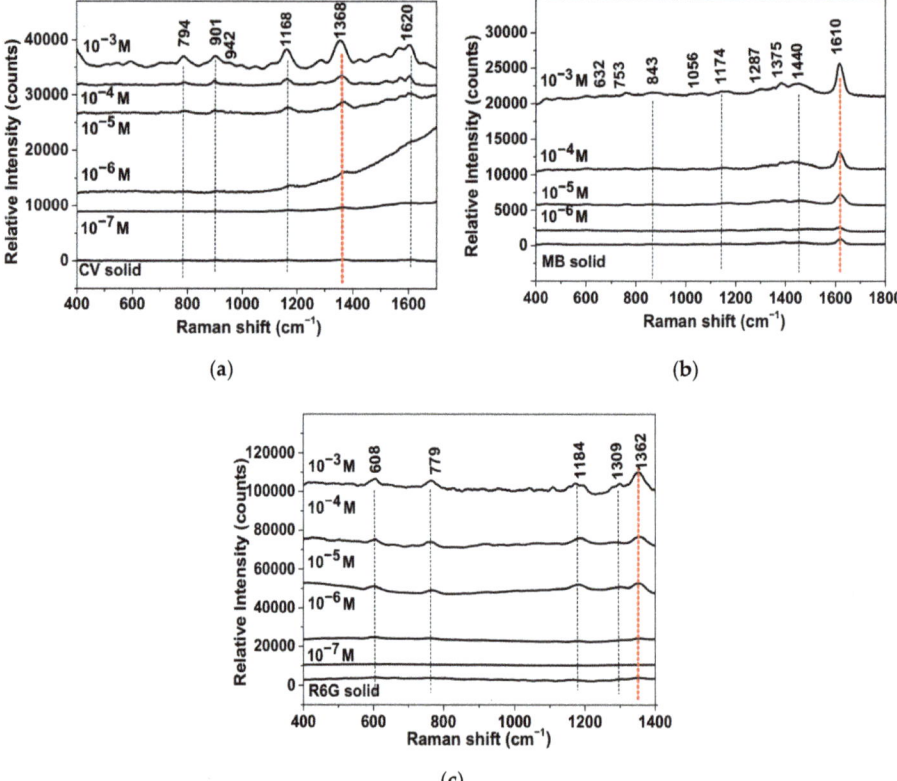

Figure 6. Raman spectra of (**a**) CV, (**b**) MB, and (**c**) R6G powders compared with SERS/SERRS spectra of the analytes with different concentrations decorated on a Cu/g-CN substrate. All spectra were plotted following baseline subtraction (see Figure S1). The red line indicates the characteristic peak used to determine the SERS activity (see text and Supporting Information). The spectra have been vertically displaced for clarity. The counts shown correspond to an integration time of 100 s of 50 accumulated scans, each with a 2-s exposure period. We note that for the CV sample prepared with a 10^{-7} M concentration, an additional fluorescence background contribution remained that was not fully accounted for by the baseline subtraction procedure. This was most likely due to some analyte molecules that were not completely removed from the underlying glass support by the washing procedure.

Because CV has a main UV-visible absorption band extending between 420–640 nm with its principal maximum near 580 nm, resonance Raman (RR) effects were expected to occur for the bulk crystalline sample using 532 nm excitation [81]. However, using the acquisition conditions applied here, we could barely observe the vibrational peaks of CV rising above the background when presented at the same vertical expansion as the SERS/SERRS spectra (Figure 6a). For the analyte deposited on the Cu/gCN support at 10^{-3} M concentration, we clearly observed the spectrum of CV with peaks at 794, 901, 942, 1168, 1368, and 1620 cm^{-1} (Figure 6a). Although the positions and relative intensities generally resembled those of published spectra of the solid sample obtained with different laser wavelengths [81], we observed the most significant enhancement in the intensity of the feature at 1368 cm^{-1} that contains a main contribution from symmetric ϕ –N (ϕ = phenyl) stretching [81]. As the concentration of adsorbed CV molecules was reduced by decadic amounts, the main peaks could still be clearly distinguished from the background down to 10^{-7} M (Figure 6a).

MB exhibited a UV-visible absorption that had barely begun at the 532 nm laser wavelength, and so it was expected that any RR or pre-resonance effects would be minimal for the bulk sample, while the background fluorescence should be low [82]. That was consistent with our data for the solid sample, for which the principal peak recorded at ≈1610 cm^{-1} in our study was clearly observed for the counting conditions used here (Figure 6b). However, the Cu NP SPR occurred at 580 nm, which lay within the electronic absorption profile of the chromophore, and so the SERS enhancement effects could also include SERRS. We observed SERS-active Raman peaks at 632, 753, 843, 1056, 1174, 1287, 1375, 1440, and 1610 cm^{-1} for our sample prepared from a 10^{-3} M analyte solution, with a relative intensity pattern that resembled that obtained by Anastasopolous et al. for MB molecules at 10^{-4} M concentration adsorbed on Ag NPs and excited using a 514.5 nm excitation [82]. We can clearly observe the dominant 1610 cm^{-1} peak down to a 10^{-6} M concentration, with a relative intensity that was only slightly lower than that for the bulk solid (Figure 6b).

R6G exhibited an electronic absorption maximum at 530 nm that matched the laser excitation wavelength and also lay close to the SPR of Cu NPs, and so RR effects were expected to be observed for both the bulk solid and adsorbed molecules in our study using a 532-nm laser excitation. As observed by previous researchers, the SERRS effect on Au nanoparticles or roughened Cu surfaces resulted in the appearance of the characteristic Raman peaks of the adsorbed molecules using red (632.8 nm) to blue (488 nm) excitation lines [83,84]. Our data show that the spectrum could be detected down to at least 10^{-6} M, where the relative intensities were comparable to those of the bulk solid (Figure 6c). Overall, our results demonstrated that the Cu/g-CN hybrid substrate exhibited good SERS/SERRS sensitivity for the CV, MB, and R6G molecular analytes.

3.3. Estimation of SERS/SERRS Enhancement Factors

In order to further investigate the SERS/SERRS enhancement activities of the Cu/gCN substrate, the surface was decorated with a sufficiently high concentration of analyte molecules (CV, MB, and R6G, 10^{-3} M) to ensure full a monolayer coverage of the NP surface. We assumed that the probe molecules were uniformly adsorbed on the support surfaces. The relative intensities of characteristic peaks for the three molecules were then compared with those obtained from a dried spot (2 µL; ≈10 mm^2) of pure analyte solution prepared at the same concentration (Figure 7a–c). The peaks at 1368, 1610, and 1362 cm^{-1} were considered for CV, MB, and R6G analytes, respectively. SERS/SERRS enhancement factors (EFs) were then estimated using the following equation [85–87]:

$$EF = (I_{SERS}/n_{SERS})/(I_{Raman}/n_{Raman}) \qquad (1)$$

Here, n_{Raman} and n_{SERS} were the number of analyte molecules included within the laser beam sampling volume for the normal Raman and SERS samples, respectively. I_{SFRS} and I_{Raman} were the absolute intensity values measured for the same mode in the Raman and SERS/SERRS spectra of the analyte. The area of the laser illumination (A_{laser}) constituted a circle ≈30 µm in radius (r_{laser}). The focused laser beam penetrated ≈4 µm into the material [86,88]. We considered the Cu NPs to be spherical with a radius (r_{NP}) of ≈16 nm (estimated from the XRD and TEM results), and that the particles were half-way embedded into the gCN matrix. We neglected spaces between the NPs.

The number of analyte molecules illuminated by the laser beam for the solid and SERS samples were then obtained using:

$$n_{SERS} = (n_{NPs} \cdot A_{NP})/(A_{Analyte}) \quad (2)$$

Here, n_{NPs}, A_{NP}, and $A_{Analyte}$ were the number of NPs in the area illuminated by the laser, the surface area per NP, and the analyte molecule surface area, respectively. n_{Raman} was then obtained using the following relation:

$$n_{Raman} = (N_A \cdot V_{Analyte} \cdot d_{Analyte})/(M_{Analyte}) \quad (3)$$

where N_A is Avogadro's number, and $V_{Analyte}$, $d_{Analyte}$, and $M_{Analyte}$ were the effective volume of analyte in the laser beam, the density, and molecular weights of the related analyte (Table 2), respectively. The *EFs* were then estimated using Equation (1) for the different analytes (Table 2). The determined values were ≈7.2 × 10^7, 2.3 × 10^7, and 1.3 × 10^7 for the CV, MB, and R6G analytes on the Cu/gCN substrate, respectively (see Method S1 for further details).

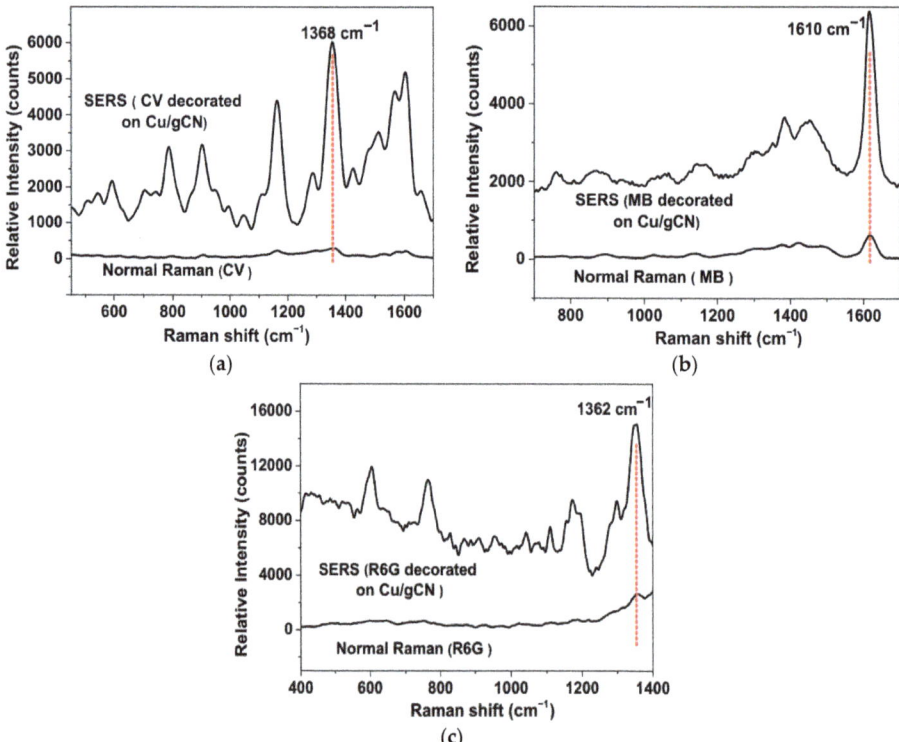

Figure 7. Comparison of the "normal" (i.e., non-SERS) Raman spectra of (**a**) CV, (**b**) MB, and (**c**) R6G with the SERS/SERRS spectra of the three analytes decorated on a Cu/gCN support. The concentration of the analytes was 10^{-3} M for recording both the "normal" Raman and SERS spectra. Spectra were plotted following subtraction of the fluorescence background from the organic molecules (see Figure S1). The red line indicates the characteristic peak used to determine the SERS activity (see text and Supporting Information). The spectra have been vertically displaced for clarity. The counts represent an integration time of 100 s of 50 accumulated scans, each with a 2 s exposure period in all cases.

Table 2. Physical characteristics of the analytes used to estimate *EFs*.

Analyte	Density (g/mL)	Surface Area (nm^2)	n_{Raman}/n_{SERS}	I_{SERS}/I_{Raman}	Cu/g-CN *EFs*
Crystal violet	1	1.2	$5.4 \times 10^{15}\pi/1.5 \times 10^9\pi$	20	7.2×10^7
Methylene blue	0.98	1.3	$6.6 \times 10^{15}\pi/1.4 \times 10^9\pi$	6.6	2.3×10^7
Rhodamine 6G	1.26	0.6	$6 \times 10^{15}\pi/3 \times 10^9\pi$	6.4	1.3×10^7

Because of the several assumptions and approximations used to obtain these *EFs*, they can only be considered as guide values for comparison with SERS and SERRS *EF* values recorded for these and related analytes on noble metal nanoparticles supported on semiconducting surfaces. Su et al. prepared a Au/MoS$_2$ nanocomposite using in situ growth of Au NPs on MoS$_2$ nanosheet surfaces, and demonstrated that the Au@MoS$_2$ substrates exhibited an 8.2×10^5 enhancement for the 1362 cm^{-1} Raman mode of R6G [89]. Jiang et al. reported that g-C$_3$N$_4$/Ag NPs exhibited *EFs* of 4.6×10^8 and 2.1×10^9 using CV as an analyte [70,71]. These authors linked the large *EF* values to multiple enhancement contributions, and that charge transfer between the g-C$_3$N$_4$ and the metal surface protected the Ag NPs from oxidation. There are only a few literature reports on the use of pure Cu NPs as a SERS substrate, either supported or in colloidal suspension [16,23–25]. Dendisová-Vyškovská et al. found $EF = 1.7 \times 10^5$ for the 1590 cm^{-1} band of 4-aminobenzenethiol (4-ABT) [16], and Ding et al. obtained an $EF = 1.6 \times 10^7$ for CV adsorbed on an array of Cu NPs deposited using sputtering on the surface of a monolayered colloidal crystal of 350 nm polystyrene spheres [23]. The SERS/SERRS results varied dramatically as a function of NP size controlled by the deposition time, with the most intense spectra recorded for an 18 min deposition with particles ≈150–200 nm in size. Although the exciting wavelength (532 nm) was the same as that used in our study, the SERS/SERRS spectrum recorded for CV was slightly different to our result in that the 1619 cm^{-1} band was much more intense. However, the fact that the *EFs* obtained in both studies is comparable is encouraging.

4. Conclusions

Our results confirm that Cu/gCN nanomaterial composites can be useful for the development of SERS/SERRS applications. Our Cu NPs on sp^2-bonded carbon nitride supports were fabricated using pulsed laser ablation of a Cu plate immersed in acetonitrile solvent. We carried out chemical, structural, and spectroscopic analyses to study the nature of the Cu NPs and their gCN support. The N:C ratio of the gCN material determined using bulk analysis was ≈1.3, significantly lower than the semiconducting carbon nitride compounds based on heptazine- to triazine-based structures, but it contained close-to-layered gCN materials containing mixed C–N/C–C bonding prepared using vapor deposition techniques. Such materials might contain locally electronically conducting domains. The degree of layer polymerization was incomplete, as shown by strong IR and Raman signals from terminal C≡N bonds. The Cu/gCN hybrid nanocomposites exhibited strong visible absorption extending toward the near-IR region with an SPR signal from the metal NPs at 580 nm. SERS/SERRS enhancement activity was tested using three analyte molecules (CV, MB, and R6G) prepared with initial solution concentrations of 10^{-3}–10^{-7} mol/L in acetonitrile and adsorbed onto the Cu/gCN nanocomposite surfaces. The observed SERS/SERRS enhancement factors were on the order of 10^7, comparable with *EF* values found for nanoparticle arrays of Cu produced using sputtering onto templates and other noble metal NPs supported by gCN materials. Further studies are now needed to investigate the possibility of cooperative interactions between the gCN support and the metal NPs that could enhance the optoelectronic effects leading to SERS/SERRS enhancement, as well as in stabilizing the Cu NPs.

Supplementary Materials: The Supplementary Materials are available online at http://www.mdpi.com/2079-4991/9/9/1223/s1.

Author Contributions: The study was devised by R.M. and synthesis, Raman, and SERS experiments were performed by H.D.-A. TEM and XPS characterization was performed by T.S.M. All authors participated in the interpretation of the results. H.D.-A. wrote a first draft of the manuscript that was completed and edited by P.F.M. and T.S.M.

Funding: R.M. thanks the Iran Nanotechnology Initiative Council (INIC) for financial support of the pulsed laser ablation setup. P.F.M. acknowledges support from the EU Graphene Flagship under Horizon 2020 Research and Innovation program grant agreement No. 696656—Graphene Core2. T.S.M. thanks the UK Engineering Physical Research Council for support via the EPSRC Postdoctoral Fellowship EP/P023851/1.

Conflicts of Interest: The authors declare no conflict of interest. The funders had no role in the design of the study; in the collection, analyses, or interpretation of data; in the writing of the manuscript, or in the decision to publish the results.

References

1. Bell, S.E.J.; Sirimuthu, N.M.S. Surface-Enhanced Raman Spectroscopy (SERS) for Sub-Micromolar Detection of DNA/RNA Mononucleotides. *J. Am. Chem. Soc.* **2006**, *128*, 15580–15581. [CrossRef] [PubMed]
2. Ambrosio, R.C.; Gewirth, A.A. Characterization of Water Structure on Silver Electrode Surfaces by SERS with Two-Dimensional Correlation Spectroscopy. *Anal. Chem.* **2010**, *82*, 1305–1310. [CrossRef] [PubMed]
3. Roguska, A.; Kudelski, A.; Pisarek, M.; Lewandowska, M.; Kurzydłowski, K.; Janik-Czachor, M. In situ spectroelectrochemical surface-enhanced Raman scattering (SERS) investigations on composite Ag/TiO$_2$-nanotubes/Ti substrates. *Surf. Sci.* **2009**, *603*, 2820–2824. [CrossRef]
4. Qiu, C.; Zhang, L.; Wang, H.; Jiang, C. Surface-Enhanced Raman Scattering on Hierarchical Porous Cuprous Oxide Nanostructures in Nanoshell and Thin-Film Geometries. *J. Phys. Chem. Lett.* **2012**, *3*, 651–657. [CrossRef] [PubMed]
5. Fateixa, S.; Nogueira, H.I.S.; Trindade, T.; Caria, S.F.; Nogueira, H.S. Hybrid nanostructures for SERS: Materials development and chemical detection. *Phys. Chem. Chem. Phys.* **2015**, *17*, 21046–21071. [CrossRef] [PubMed]
6. Han, X.X.; Ji, W.; Zhao, B.; Ozaki, Y. Semiconductor-enhanced Raman scattering: Active nanomaterials and applications. *Nanoscale* **2017**, *9*, 4847–4861. [CrossRef] [PubMed]
7. Lombardi, J.R.; Birke, R.L. A Unified Approach to Surface-Enhanced Raman Spectroscopy. *J. Phys. Chem. C* **2008**, *112*, 5605–5617. [CrossRef]
8. Kneipp, K.; Wang, Y.; Kneipp, H.; Perelman, L.T.; Itzkan, I.; Dasari, R.R.; Feld, M.S. Single Molecule Detection Using Surface-Enhanced Raman Scattering (SERS). *Phys. Rev. Lett.* **1997**, *78*, 1667–1670. [CrossRef]
9. Nie, S.; Emory, S.R. Probing single molecules and single nanoparticles by SERS. *Science* **1997**, *275*, 1102–1106. [CrossRef] [PubMed]
10. Xu, H.; Bjerneld, E.J.; Käll, M.; Börjesson, L. Spectroscopy of Single Hemoglobin Molecules by Surface Enhanced Raman Scattering. *Phys. Rev. Lett.* **1999**, *83*, 4357–4360. [CrossRef]
11. Schlücker, S. Surface-Enhanced Raman Spectroscopy: Concepts and Chemical Applications. *Angew. Chem. Int. Ed.* **2014**, *53*, 4756–4795. [CrossRef] [PubMed]
12. Wang, X.; Shi, W.; She, G.; Mu, L. Using Si and Ge Nanostructures as Substrates for Surface-Enhanced Raman Scattering Based on Photoinduced Charge Transfer Mechanism. *J. Am. Chem. Soc.* **2011**, *133*, 16518–16523. [CrossRef] [PubMed]
13. Ji, W.; Zhao, B.; Ozaki, Y. Semiconductor materials in analytical applications of surface-enhanced Raman scattering. *J. Raman Spectrosc.* **2016**, *47*, 51–58. [CrossRef]
14. Musumeci, A.; Gosztola, D.; Schiller, T.; Dimitrijevic, N.M.; Mujica, V.; Martin, D.; Rajh, T. SERS of Semiconducting Nanoparticles (TiO$_2$ Hybrid Composites). *J. Am. Chem. Soc.* **2009**, *131*, 6040–6041. [CrossRef] [PubMed]
15. Teguh, J.S.; Xing, B.; Yeow, E.K.L.; Liu, F.; Xing, P.D.B.; Yeow, P.D.E.K.L. Surface-Enhanced Raman Scattering (SERS) of Nitrothiophenol Isomers Chemisorbed on TiO$_2$. *Chem. Asian J.* **2012**, *7*, 975–981. [CrossRef] [PubMed]

16. Dendisova-Vyskovska, M.; Prokopec, V.; Člupek, M.; Matějka, P. Comparison of SERS effectiveness of copper substrates prepared by different methods: What are the values of enhancement factors? *J. Raman Spectrosc.* **2012**, *43*, 181–186. [CrossRef]
17. McNay, G.; Eustace, D.; Smith, W.E.; Faulds, K.; Graham, D. Surface-enhanced Raman scattering (SERS) and resonance Raman scattering (SERRS): A review of applications. *Appl. Spectrosc.* **2011**, *65*, 825–837. [CrossRef] [PubMed]
18. Bell, S.E.J.; McCourt, M.R. SERS enhancement by aggregated Au colloids: Effect of particle size. *Phys. Chem. Chem. Phys.* **2009**, *11*, 7455–7462. [CrossRef] [PubMed]
19. Jung, M.; Kim, S.K.; Lee, S.; Kim, J.H.; Woo, D. Ag nanodot array as a platform for surface-enhanced Raman scattering. *J. Nanophotonics* **2013**, *7*, 073798. [CrossRef]
20. Jung, G.B.; Bae, Y.M.; Lee, Y.J.; Ryu, S.H.; Park, H.-K. Nanoplasmonic Au nanodot arrays as an SERS substrate for biomedical applications. *Appl. Surf. Sci.* **2013**, *282*, 161–164. [CrossRef]
21. Ameer, F.S.; Pittman, C.U.; Zhang, D. Quantification of Resonance Raman Enhancement Factors for Rhodamine 6G (R6G) in Water and on Gold and Silver Nanoparticles: Implications for Single-Molecule R6G SERS. *J. Phys. Chem. C* **2013**, *117*, 27096–27104. [CrossRef]
22. Roguska, A.; Kudelski, A.; Pisarek, M.; Opara, M.; Janik-Czachor, M. Surface-enhanced Raman scattering (SERS) activity of Ag, Au and Cu nanoclusters on TiO_2-nanotubes/Ti substrate. *Appl. Surf. Sci.* **2011**, *257*, 8182–8189. [CrossRef]
23. Ding, Q.; Hang, L.; Ma, L. Controlled synthesis of Cu nanoparticle arrays with surface enhanced Raman scattering effect performance. *RSC Adv.* **2018**, *8*, 1753–1757. [CrossRef]
24. Nguyen, T.B.; Vu, T.K.T.; Nguyen, Q.D.; Nguyen, T.D.; Nguyen, T.A.; Trinh, T.H. Preparation of metal nanoparticles for surface enhanced Raman scattering by laser ablation method. *Adv. Nat. Sci. Nanosci. Nanotechnol.* **2012**, *3*, 025016. [CrossRef]
25. Muniz-Miranda, M.; Gellini, C.; Giorgetti, E. Surface-Enhanced Raman Scattering from Copper Nanoparticles Obtained by Laser Ablation. *J. Phys. Chem. C* **2011**, *115*, 5021–5027. [CrossRef]
26. Khan, A.; Rashid, A.; Younas, R.; Chong, R. A chemical reduction approach to the synthesis of copper nanoparticles. *Int. Nano Lett.* **2016**, *6*, 21–26. [CrossRef]
27. Soomro, R.A.; Sirajuddin, S.T.H.S.; Memon, N.; Shah, M.R.; Kalwar, N.H.; Hallam, K.R.; Shah, A. Synthesis of Air Stable Copper Nanoparticles and Their Use in Catalysis. *Adv. Mater. Lett.* **2014**, *5*, 191–198. [CrossRef]
28. Dang, T.M.D.; Le, T.T.T.; Fribourg-Blanc, E.; Dang, M.C. Synthesis and optical properties of copper nanoparticles prepared by a chemical reduction method. *Adv. Nat. Sci. Nanosci. Nanotechnol.* **2011**, *2*, 015009. [CrossRef]
29. Matikainen, A.; Nuutinen, T.; Itkonen, T.; Heinilehto, S.; Puustinen, J.; Hiltunen, J.; Lappalainen, J.; Karioja, P.; Vahimaa, P. Atmospheric oxidation and carbon contamination of silver and its effect on surface-enhanced Raman spectroscopy (SERS). *Sci. Rep.* **2016**, *6*, 37192. [CrossRef]
30. Zeng, F.; Xu, D.; Zhan, C.; Liang, C.; Zhao, W.; Zhang, J.; Feng, H.; Ma, X. Surfactant-Free Synthesis of Graphene Oxide Coated Silver Nanoparticles for SERS Biosensing and Intracellular Drug Delivery. *ACS Appl. Nano Mater.* **2018**, *1*, 2748–2753. [CrossRef]
31. Neddersen, J.; Chumanov, G.; Cotton, T.M. Laser Ablation of Metals: A New Method for Preparing SERS Active Colloids. *Appl. Spectrosc.* **1993**, *47*, 1959–1964. [CrossRef]
32. Gondal, M.A.; Qahtan, T.F.; Dastageer, M.A.; Maganda, Y.W.; Anjum, D.H. Synthesis of Cu/Cu_2O nanoparticles by laser ablation in deionized water and their annealing transformation into CuO nanoparticles. *J. Nanosci. Nanotechnol.* **2013**, *13*, 5759–5766. [CrossRef] [PubMed]
33. Aghdam, H.D.; Azadi, H.; Esmaeilzadeh, M.; Bellah, S.M.; Malekfar, R. Ablation time and laser fluence impacts on the composition, morphology and optical properties of copper oxide nanoparticles. *Opt. Mater.* **2019**, *91*, 433–438. [CrossRef]
34. Abdulateef, S.A.; MatJafri, M.Z.; Omar, A.F.; Ahmed, N.M.; Azzez, S.A.; Ibrahim, I.M.; Al-Jumaili, B.E.B. Preparation of CuO nanoparticles by laser ablation in liquid. *AIP Conf. Proc.* **2016**, *1733*, 020035.
35. Kudelski, A.; Grochala, W.; Janik-Czachor, M.; Bukowska, J.; Szummer, A.; Dolata, M. Surface-enhanced Raman scattering (SERS) at Copper(I) oxide. *J. Raman Spectrosc.* **1998**, *29*, 431–435. [CrossRef]
36. Amendola, V.; Rizzi, G.A.; Polizzi, S.; Meneghetti, M. Synthesis of Gold Nanoparticles by Laser Ablation in Toluene: Quenching and Recovery of the Surface Plasmon Absorption. *J. Phys. Chem. B* **2005**, *109*, 23125–23128. [CrossRef] [PubMed]

37. Nikov, R.G.; Nedyalkov, N.N.; Atanasov, P.A.; Karashanova, D.B. Laser-assisted fabrication and size distribution modification of colloidal gold nanostructures by nanosecond laser ablation in different liquids. *Appl. Phys. A* **2017**, *123*, 490. [CrossRef]
38. Cho, H.-G. Matrix Infrared Spectra and DFT Computations of CH_2CNH and CH_2NCH Produced from CH_3CN by Laser-Ablation Plume Radiation. *Bull. Korean Chem. Soc.* **2013**, *34*, 1361–1365. [CrossRef]
39. Andrews, L.; Citra, A. Infrared Spectra and Density Functional Theory Calculations on Transition Metal Nitrosyls. Vibrational Frequencies of Unsaturated Transition Metal Nitrosyls. *Chem. Rev.* **2002**, *102*, 885–912. [CrossRef]
40. Cho, H.-G.; Andrews, L. Matrix Infrared Spectra and Density Functional Calculations of the H_2CCN and H_2CNC Radicals Produced from CH_3CN. *J. Phys. Chem. A* **2011**, *115*, 8638–8642. [CrossRef]
41. Amendola, V.; Meneghetti, M. What controls the composition and the structure of nanomaterials generated by laser ablation in liquid solution? *Phys. Chem. Chem. Phys.* **2013**, *15*, 3027–3046. [CrossRef] [PubMed]
42. Hamad, S.; Podagatlapalli, G.K.; Rao, S.V.; Syed, H.; Soma, V.R. Explosives Detection with Copper Nanostructures Fabricated using Ultrafast Laser Ablation in Acetonitrile. *Light Energy Environ.* **2014**. [CrossRef]
43. Kessler, F.K.; Zheng, Y.; Schwarz, D.; Merschjann, C.; Schnick, W.; Wang, X.; Bojdys, M.J. Functional carbon nitride materials—Design strategies for electrochemical devices. *Nat. Rev. Mater.* **2017**, *2*, 17030. [CrossRef]
44. Miller, T.; Jorge, A.B.; Suter, T.M.; Sella, A.; Corà, F.; McMillan, P.F. Carbon nitrides: Synthesis and characterization of a new class of functional materials. *Phys. Chem. Chem. Phys.* **2017**, *19*, 15613–15638. [CrossRef] [PubMed]
45. Wen, J.; Xie, J.; Chen, X.; Li, X. A review on g-C 3 N 4 -based photocatalysts. *Appl. Surf. Sci.* **2017**, *391*, 72–123. [CrossRef]
46. Mansor, N.; Miller, T.S.; Dedigama, I.; Jorge, A.B.; Jia, J.; Brázdová, V.; Mattevi, C.; Gibbs, C.; Hodgson, D.; Shearing, P.R.; et al. Graphitic Carbon Nitride as a Catalyst Support in Fuel Cells and Electrolyzers. *Electrochim. Acta* **2016**, *222*, 44–57. [CrossRef]
47. Algara-Siller, G.; Severin, N.; Chong, S.Y.; Björkman, T.; Palgrave, R.G.; Laybourn, A.; Antonietti, M.; Khimyak, Y.Z.; Krasheninnikov, A.V.; Rabe, J.P.; et al. Triazine-Based Graphitic Carbon Nitride: A Two-Dimensional Semiconductor. *Angew. Chem.* **2014**, *126*, 7580–7585. [CrossRef]
48. Wirnhier, E.; Doblinger, M.; Gunzelmann, D.; Senker, J.; Lotsch, B.V.; Schnick, W. Poly(triazine imide) with intercalation of lithium and chloride ions [(C3N3)2(NHxLi1-x)3•LiCl]: A crystalline 2D carbon nitride network. *Chem. Eur. J.* **2011**, *17*, 3213–3221. [CrossRef]
49. Ladva, S.A.; Travis, W.; Quesada-Cabrera, R.; Rosillo-Lopez, M.; Afandi, A.; Li, Y.; Jackman, R.B.; Bear, J.C.; Parkin, I.P.; Blackman, C.; et al. Nanoscale, conformal films of graphitic carbon nitride deposited at room temperature: A method for construction of heterojunction devices. *Nanoscale* **2017**, *9*, 16586–16760. [CrossRef]
50. Suter, T.M.; Brazdova, V.; McColl, K.; Miller, T.S.; Nagashima, H.; Salvadori, E.; Sella, A.; Howard, C.A.; Kay, C.W.M.; Cora, F.; et al. Synthesis, Structure and Electronic Properties of Graphitic Carbon Nitride Films. *J. Phys. Chem. C* **2018**, *122*, 25183–25194. [CrossRef]
51. Zhang, Z.J.; Fan, S.; Lieber, C.M. Growth and composition of covalent carbon nitride solids. *Appl. Phys. Lett.* **1995**, *66*, 3582–3584. [CrossRef]
52. Rodil, S.E.; Muhl, S.; Maca, S.; Ferrari, A. Optical gap in carbon nitride films. *Thin Solid Films* **2003**, *433*, 119–125. [CrossRef]
53. Inagaki, M.; Toyoda, M.; Soneda, Y.; Morishita, T. Nitrogen-doped carbon materials. *Carbon* **2018**, *132*, 104–140. [CrossRef]
54. González, P.; Soto, R.; Parada, E.; Redondas, X.; Chiussi, S.; Serra, J.; Pou, J.; Leon, B.; Pérez-Amor, M. Carbon nitride films prepared by excimer laser ablation. *Appl. Surf. Sci.* **1997**, *109*, 380–383. [CrossRef]
55. Bulíř, J.; Jelínek, M.; Vorlicek, V.; Zemek, J.; Perina, V. Study of nitrogen pressure effect on the laser-deposited amorphous carbon films. *Thin Solid Films* **1997**, *292*, 318–323. [CrossRef]
56. Zhao, X.; Ong, C.W.; Tsang, Y.C.; Wong, Y.W.; Chan, P.W.; Choy, C.L. Reactive pulsed laser deposition of CNx films. *Appl. Phys. Lett.* **1995**, *66*, 2652–2654. [CrossRef]
57. Ong, C.W.; Zhao, X.-A.; Tsang, Y.; Choy, C.; Chan, P. Effects of substrate temperature on the structure and properties of reactive pulsed laser deposited CNx films. *Thin Solid Films* **1996**, *280*, 1–4. [CrossRef]
58. Alexandrescu, R.; Huisken, F.; Pugna, G.; Crunteanu, A.; Petcu, S.; Cojocaru, S.; Cireasa, R.; Morjan, I.; Cojocaru, C.S. Preparation of carbon nitride fine powder by laser induced gas-phase reactions. *Appl. Phys. A* **1997**, *65*, 207–213. [CrossRef]

59. Sharma, A.K.; Ayyub, P.; Multani, M.S.; Adhi, K.P.; Ogale, S.B.; Sunderaraman, M.; Upadhyay, D.D.; Banerjee, S. Synthesis of crystalline carbon nitride thin films by laser processing at a liquid–solid interface. *Appl. Phys. Lett.* **1996**, *69*, 3489–3491. [CrossRef]
60. Yang, L.; May, P.W.; Yin, L.; Brown, R.; Scott, T.B. Direct Growth of Highly Organized Crystalline Carbon Nitride from Liquid-Phase Pulsed Laser Ablation. *Chem. Mater.* **2006**, *18*, 5058–5064. [CrossRef]
61. Li, H.; Xu, Y.; Sitinamaluwa, H.; Wasalathilake, K.; Galpaya, D.; Yan, C. Cu nanoparticles supported on graphitic carbon nitride as an efficient electrocatalyst for oxygen reduction reaction. *Chin. J. Catal.* **2017**, *38*, 1006–1010. [CrossRef]
62. Le, S.; Fang, B.; Jiang, T.; Zhao, Q.; Liu, X.; Li, Y.; Gong, M. Cu-doped mesoporous graphitic carbon nitride for enhanced visible-light driven photocatalysis. *RSC Adv.* **2016**, *6*, 38811–38819. [CrossRef]
63. Muniandy, L.; Adam, F.; Mohamed, A.R.; Iqbal, A.; Rahman, N.R.A. Cu^{2+} coordinated graphitic carbon nitride (Cu-g-C_3N_4) nanosheets from melamine for the liquid phase hydroxylation of benzene and VOCs. *Appl. Surf. Sci.* **2017**, *398*, 43–55. [CrossRef]
64. Gao, J.; Wang, J.; Qian, X.; Dong, Y.; Xu, H.; Song, R.; Yan, C.; Zhu, H.; Zhong, Q.; Qian, G.; et al. One-pot synthesis of copper-doped graphitic carbon nitride nanosheet by heating Cu–melamine supramolecular network and its enhanced visible-light-driven photocatalysis. *J. Solid State Chem.* **2015**, *228*, 60–64. [CrossRef]
65. Tahir, B.; Tahir, M.; Amin, N.A.S. Photo-induced CO_2 reduction by CH_4/H_2O to fuels over Cu-modified g-C_3N_4 nanorods under simulated solar energy. *Appl. Surf. Sci.* **2017**, *419*, 875–885. [CrossRef]
66. Shi, G.; Yang, L.; Liu, Z.; Chen, X.; Zhou, J.; Yu, Y. Photocatalytic reduction of CO_2 to CO over copper decorated g-C_3N_4 nanosheets with enhanced yield and selectivity. *Appl. Surf. Sci.* **2018**, *427*, 1165–1173. [CrossRef]
67. Wolosiuk, A.; Tognalli, N.G.; Martínez, E.D.; Granada, M.; Fuertes, M.C.; Troiani, H.; Bilmes, S.A.; Fainstein, A.; Soler-Illia, G.J.A.A. Silver Nanoparticle-Mesoporous Oxide Nanocomposite Thin Films: A Platform for Spatially Homogeneous SERS-Active Substrates with Enhanced Stability. *ACS Appl. Mater. Interfaces* **2014**, *6*, 5263–5272. [CrossRef] [PubMed]
68. Yang, K.-H.; Liu, Y.-C.; Hsu, T.-C.; Juang, M.-Y. Strategy to improve stability of surface-enhanced raman scattering-active Ag substrates. *J. Mater. Chem.* **2010**, *20*, 7530–7535. [CrossRef]
69. Mai, F.-D.; Yang, K.-H.; Liu, Y.-C.; Hsu, T.-C. Improved stabilities on surface-enhanced Raman scattering-active Ag/Al_2O_3 films on substrates. *Analyst* **2012**, *137*, 5906–5912. [CrossRef] [PubMed]
70. Jiang, J.; Zhu, L.; Zou, J.; Ou-Yang, L.; Zheng, A.; Tang, H. Micro/nano-structured graphitic carbon nitride—Ag nanoparticle hybrids as surface-enhanced Raman scattering substrates with much improved long-term stability. *Carbon* **2015**, *87*, 193–205. [CrossRef]
71. Jiang, J.; Zou, J.; Wee, A.T.S.; Zhang, W. Use of Single-Layer g-C_3N_4/Ag Hybrids for Surface-Enhanced Raman Scattering (SERS). *Sci. Rep.* **2016**, *6*, 34599. [CrossRef] [PubMed]
72. Schneider, C.A.; Rasband, W.S.; Eliceiri, K.W. NIH Image to ImageJ: 25 years of Image Analysis. *Nat. Methods* **2012**, *9*, 671–675. [CrossRef] [PubMed]
73. Biesinger, M.C.; Payne, B.P.; Grosvenor, A.P.; Lau, L.W.; Gerson, A.R.; Smart, R.S. Resolving surface chemical states in XPS analysis of first row transition metals, oxides and hydroxides: Cr, Mn, Fe, Co and Ni. *Appl. Surf. Sci.* **2011**, *257*, 2717–2730. [CrossRef]
74. Liu, P.; Wang, H.; Li, X.; Rui, M.; Zeng, H. Localized surface plasmon resonance of Cu nanoparticles by laser ablation in liquid media. *RSC Adv.* **2015**, *5*, 79738–79745. [CrossRef]
75. Zambon, A.; Mouesca, J.M.; Gheorghiu, C.; Bayle, P.A.; Pécaut, J.; Claeys-Bruno, M.; Gambarelli, S.; Dubois, L. s-Heptazine oligomers: Promising structural models for graphitic carbon nitride. *Chem. Sci.* **2016**, *7*, 945–950. [CrossRef] [PubMed]
76. Fina, F.; Callear, S.K.; Carins, G.M.; Irvine, J.T.S. Structural Investigation of Graphitic Carbon Nitride via XRD and Neutron Diffraction. *Chem. Mater.* **2015**, *27*, 2612–2618. [CrossRef]
77. Tyborski, T.; Merschjann, C.; Orthmann, S.; Yang, F.; Lux-Steiner, M.-C.; Schedel-Niedrig, T. Crystal structure of polymeric carbon nitride and the determination of its process-temperature-induced modifications. *J. Phys. Condens. Matter* **2013**, *25*, 395402. [CrossRef]
78. Bonner, O.; Curry, J.D. Infrared spectra of liquid H_2O and D_2O. *Infrared Phys.* **1970**, *10*, 91–94. [CrossRef]
79. Rodil, S.E.; Ferrari, A.; Robertson, J.; Muhl, S. Infrared spectra of carbon nitride films. *Thin Solid Films* **2002**, *420*, 122–131. [CrossRef]

80. Ferrari, A.C.; Rodil, S.E.; Robertson, J. Interpretation of infrared and Raman spectra of amorphous carbon nitrides. *Phys. Rev. B* **2003**, *67*, 155306. [CrossRef]
81. Angeloni, L.; Smulevich, G.; Marzocchi, M.P. Resonance Raman spectrum of crystal violet. *J. Raman Spectrosc.* **1979**, *8*, 305–310. [CrossRef]
82. Anastasopoulos, J.; Beobide, A.S.; Manikas, A.; Voyiatzis, G. Quantitative surface-enhanced resonance Raman scattering analysis of methylene blue using silver colloid. *J. Raman Spectrosc.* **2017**, *48*, 1762–1770. [CrossRef]
83. Mo, Y.; Lei, J.; Li, X.; Wächter, P. Surface enhanced Raman scattering of rhodamine 6G and dye 1555 adsorbed on roughened copper surfaces. *Solid State Commun.* **1988**, *66*, 127–131. [CrossRef]
84. Ujihara, M.; Dang, N.M.; Imae, T. Surface-Enhanced Resonance Raman Scattering of Rhodamine 6G in Dispersions and on Films of Confeito-Like Au Nanoparticles. *Sensors* **2017**, *17*, 2563. [CrossRef] [PubMed]
85. Le Ru, E.C.; Blackie, E.; Meyer, M.; Etchegoin, P.G. Surface Enhanced Raman Scattering Enhancement Factors: A Comprehensive Study. *J. Phys. Chem. C* **2007**, *111*, 13794–13803. [CrossRef]
86. Jiang, L.; You, T.; Yin, P.; Shang, Y.; Zhang, D.; Guo, L.; Yang, S. Surface-enhanced Raman scattering spectra of adsorbates on Cu_2O nanospheres: Charge-transfer and electromagnetic enhancement. *Nanoscale* **2013**, *5*, 2784–2789. [CrossRef] [PubMed]
87. He, S.; Chua, J.; Tan, E.K.M.; Kah, J.C.Y. Optimizing the SERS enhancement of a facile gold nanostar immobilized paper-based SERS substrate. *RSC Adv.* **2017**, *7*, 16264–16272. [CrossRef]
88. Xu, Y.; Konrad, M.P.; Trotter, J.L.; McCoy, C.P.; Bell, S.E.J. Rapid One-Pot Preparation of Large Freestanding Nanoparticle-Polymer Films. *Small* **2016**, *13*, 1602163. [CrossRef] [PubMed]
89. Su, S.; Zhang, C.; Yuwen, L.; Chao, J.; Zuo, X.; Liu, X.; Song, C.; Fan, C.; Wang, L. Creating SERS Hot Spots on MoS_2 Nanosheets with in Situ Grown Gold Nanoparticles. *ACS Appl. Mater. Interfaces* **2014**, *6*, 18735–18741. [CrossRef]

© 2019 by the authors. Licensee MDPI, Basel, Switzerland. This article is an open access article distributed under the terms and conditions of the Creative Commons Attribution (CC BY) license (http://creativecommons.org/licenses/by/4.0/).

Article

Controlling the Morphologies of Silver Aggregates by Laser-Induced Synthesis for Optimal SERS Detection

Longkun Yang [†], Jingran Yang [†], Yuanyuan Li, Pan Li, Xiaojuan Chen and Zhipeng Li *

Beijing Key Laboratory of Nano-Photonics and Nano-Structure (NPNS), Department of Physics, Capital Normal University, Beijing 100048, China; lkyang@cnu.edu.cn (L.Y.); yang.jing.ran@163.com (J.Y.); liyuanyuan1881005@163.com (Y.L.); cnulp@sina.com(P.L.); xiaojuan@buaa.edu.cn (X.C.)
* Correspondence: zpli@cnu.edu.cn
† These authors contributed equally to this work.

Received: 29 September 2019; Accepted: 22 October 2019; Published: 27 October 2019

Abstract: Controlling the synthesis of metallic nanostructures for high quality surface-enhanced Raman scattering (SERS) materials has long been a central task of nanoscience and nanotechnology. In this work, silver aggregates with different surface morphologies were controllably synthesized on a glass–solution interface via a facile laser-induced reduction method. By correlating the surface morphologies with their SERS abilities, optimal parameters (laser power and irradiation time) for SERS aggregates synthesis were obtained. Importantly, the characteristics for largest near-field enhancement were identified, which are closely packed nanorice and flake structures with abundant surface roughness. These can generate numerous hot spots with huge enhancement in nanogaps and rough surface. These results provide an understanding of the correlation between morphologies and SERS performance, and could be helpful for developing optimal and applicable SERS materials.

Keywords: silver aggregates; laser-induced synthesis; surface-enhanced Raman scattering; hot spots

1. Introduction

Surface-enhanced Raman scattering (SERS) is a powerful, nondestructive analytical tool owing to its high molecular specificity and sensitivity [1–3]. It has demonstrated promising applications in the fields of single-molecule spectroscopy [4,5], biochemical analysis [6–10], environmental monitoring [11–14], food safety [15,16], and even monitoring the reaction process at a molecular level [17–20]. The phenomenon of SERS is generally explained by a combination of electromagnetic [21–24] and chemical [25–27] enhancements. The former involves the enhancement of the electric field by the surface plasmons resonance of metallic nanoparticles. Especially, when two nanostructures are brought together, a giant local field can be generated in the gap or crevice due to the surface plasmons coupling, which is a hot spot for SERS detection [28–32]. The latter mainly originates from the charge transfer between the adsorbates and metal surface [25–27]. With respect to electromagnetic enhancement, a number of techniques have been developed to rationally design the SERS substrates with a large density of hot spots in order to improve the sensitivity and reproducibility of SERS measurements [33–39]. For instance, metallic nanostructures with various shapes, such as silver and gold spheres [40,41], cubes [42,43], polyhedrons [44,45], rods [46,47], and wires [48,49] have been chemically synthesized. When dropping the colloidal suspensions on an omniphobic or slippery substrate, SERS hot spots can be formed when the nanoparticles self-assemble during solvent evaporation [50,51]. On this slippery SERS platform, reproducibility can reach 25% for single-molecule SERS detection (~10^{-13} M) and can rapidly increase to >90% at higher detection concentrations (>10^{-12} M) [50]. On the other hand, it has been reported that silver aggregates with dense hot spots can directly grow on the interface of indium tin oxide (ITO) and reaction solution by a simple laser-induced photochemical reduction [52–55]. This laser-direct writing method provides a rapid, controllable, and low-cost way to synthesize SERS active materials. More importantly, this technique can integrate the SERS substrates directly into the microfluidic channel in a controlled fashion to create a lab-on-a-chip SERS system. It has the advantages of in situ preparation, automation, and real time detection, and avoids the unexpected contamination or oxidation degrading of SERS substrates, thus enables reproducible and sensitive SERS measurements [56–58]. With this silver aggregates-based SERS chip, the reproducibility of single-molecule SERS measurements can be raised up to ~50% [59]. We know the growth of silver aggregates on a glass–solution interface is highly dependent on laser power and irradiation time. Hence, it is critical to understand the correlation between the morphologies of aggregates and the corresponding SERS ability to optimize the performance of SERS materials synthesized by this laser-induced photochemical reduction method.

In this work, the laser-induced growths of silver aggregates on an ITO–solution interface were systematically investigated by tuning the power and irradiation time (532 nm laser). These structures can generate numerous hot spots at both the nanogaps and rough surface. By correlating the aggregates morphologies with their SERS abilities, the critical structure characteristics for large near-field enhancement were identified, which were closely packed nanorice and flake structures with abundant surface roughness. The understanding of the relation between morphology and SERS performance would be beneficial for developing optimal and applicable SERS materials.

2. Experimental Section

2.1. In Situ Synthesis of Silver Aggregates

Silver nitrate and sodium citrate dihydrate of analytical grade were bought from Sigma-Aldrich. Deionized water was used to prepare the solutions. The reactant mixture was obtained by mixing aqueous solutions of silver nitrate (0.1 mM) and sodium citrate (0.08 mM) in a 1:1 volume ratio. Then, a drip of reactant mixture was placed in a cell made up of a slide of ITO glass and a cover glass. A 532 nm continuous wave laser was focused on the ITO glass through an objective with 50× magnification (N.A. = 0.5). The power was tuned in the range of 0.4–0.9 mW by an attenuator. Laser irradiation time was set in the range of 30–180 s. The final products on ITO glass were rinsed for 5 min with deionized water to remove excess reactants. Using a scanning electron microscope (SEM, S-4800, 10 kV, Hitachi, Japan), the morphologies of silver aggregates synthesized under different power and irradiation time were characterized. With the help of coordinates on ITO glass, each characterized silver nanoaggregates could be specifically found again under an optical microscope.

2.2. SERS Measurements

The SERS measurements were performed on an inVia Renishaw Raman Spectrometer at the excitation of a 532 nm laser (Renishaw, UK). A 50× magnification (N.A. = 0.5) objective was used. Here, the laser for Raman excitation was the same as the one used for photo reduction. The Raman excitation power was about 14 µW and integration time was 10 s, unless stated otherwise. Crystal violet (CV) with a concentration of 10^{-7} M in ethanol was chosen as the SERS analyte. The SERS sample was prepared by dropping 20 µL CV solution onto the ITO slide with silver aggregates. After it dried under ambient conditions, the area of the dried spot was about 1 cm^2. For polarization measurements, the SERS spectra were repeatedly detected at the same position by changing the excitation polarization. To minimize the photobleaching-induced SERS decay [33], the integration time was set to 1 s. The intrinsic polarization dependence of the Raman instrument was calibrated by the Raman peak of silicon (111) surface.

3. Results and Discussion

Our experimental setup for the laser-induced synthesis of silver aggregates is schematically shown in Figure 1a (for details, see Experimental Section). A 532 nm continuous wave laser was focused onto the cell filled with reactant solution prepared by silver nitrate and sodium citrate. Then, the citrate reduced the silver ions to atoms at room temperature via a photooxidation mechanism, dissociating a hydrogen ion from the hydroxyl group on the citrate and converting it to acetone-1,3-dicarboxylate and carbon dioxide [60–62]. With the continuous increase of silver atoms, silver aggregate structures grew on the ITO–solution interface in a few seconds, and were observed under the microscope. The SEM image of a typical product synthesized by 60 s exposure at a laser power of 0.9 mW is shown in Figure 1b. From the SEM images, we found that the aggregate spot was about 10 μm in size and was made up of numerous nanorice and flake structures, with an average length of about 460 nm. These surface textures could form dense gaps or crevices and tips, which could generate a huge number of hot spots for SERS detection.

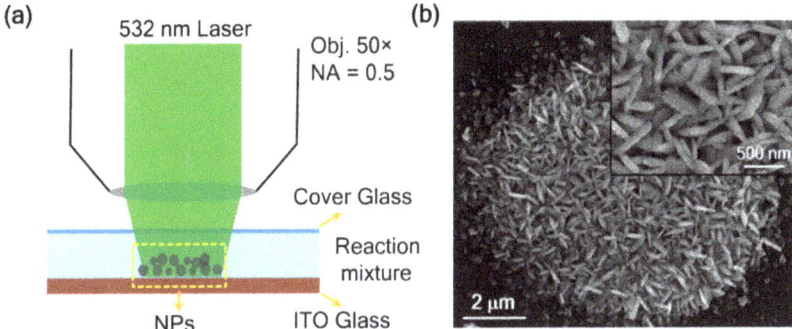

Figure 1. Laser-induced growth of silver aggregates. (**a**) Schematic of the experimental setup. (**b**) SEM image of a typical product fabricated by 60 s exposure at a laser power of 0.9 mW. Inset is a zoomed view of the silver aggregates.

The SERS performances of the prepared silver aggregates were then experimentally characterized by using 10^{-7} M CV as the probe. Curve I in Figure 2a shows the raw Raman spectrum obtained from CV powder. The Raman fingerprints at 913, 1174, 1375, 1584, and 1616 cm^{-1} were identified and were mainly from the vibrations of benzene ring [63]. The corresponding SERS spectrum of CV on the as-prepared silver aggregates is shown in curve II. By comparing to the spectrum from powder, we found that the Raman scattering intensity was greatly enhanced. According to the absorption spectrum of CV (Figure S1), resonant Raman scattering can be obtained under the excitation of 532 nm. Hence, the enhancement should come from the combined contributions of plasmonic effect and molecular resonance effect, which is surface-enhanced resonance Raman scattering. Generally, the enhancement factor (EF) can be evaluated by the following equation:

$$\mathrm{EF} = \frac{I_{SERS}}{I_{Bulk}} \times \frac{N_{Bulk}}{N_{SERS}}, \tag{1}$$

where I_{SERS} and I_{Bulk} are the intensity of a Raman mode with and without surface enhancement, respectively, and N_{SERS} and N_{Bulk} refer to the corresponding number of CV molecules [64–66]. By choosing the CV band at 1174 cm^{-1} as a reference, the EF was estimated to be 2.0×10^7 (see supporting information for details). Considering that the bulk Raman was also excited by 532 nm laser, the molecular resonance effect would be offset to some extent in EF evaluation. Hence, the EF calculated by Equation 1 can be attributed to electromagnetic enhancement of silver aggregates. Here, we should emphasize that the EF is an average value over the whole surface of silver aggregates. The enhancements on the tips of nanorices and flakes or inside the gaps of nanoaggregates would be much larger. As is known, the electromagnetic enhancement of metallic nanostructures is highly dependent on excitation polarization [67,68]. Hence, the polarization-dependent SERS of these silver aggregates were investigated. The normalized SERS intensity of CV under different incident polarizations are shown in the polar plot (Figure 2b). Unlike the highly polarization-dependent single-nanogap system [68], the aggregates structure was insensitive to the excitation polarization with the SERS intensity fluctuation at orthogonal polarizations less than 20%. This is to be expected because the as-prepared silver aggregates were made up of dense nanogaps or crevices with random sizes and orientations (Figure 1b).

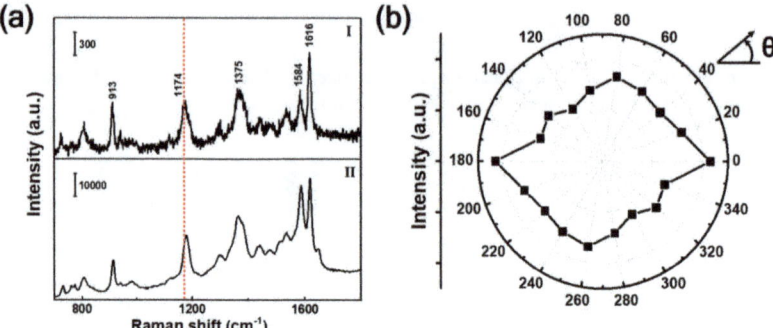

Figure 2. Surface-enhanced Raman scattering (SERS) measurements of the silver aggregates fabricated by 60 s exposure at a laser power of 0.9 mW. (**a**) Curve I: Raman spectrum of CV powder. Curve II: The SERS spectrum of CV adsorbed on the silver aggregates. (**b**) Polar plot of SERS intensity (1174 cm^{-1}) under different excitation polarizations (θ).

To seek the optimal synthesis parameters, SERS aggregates were created under different laser power (0.4, 0.6, and 0.9 mW) and irradiation time (30, 60, 120, and 180 s). The morphologies of the prepared aggregates are summarized in Figure 3a. To clarify the influence of the laser power and irradiation time, we first analyzed the morphological changes as the increase of irradiation time at certain laser power. Such as the aggregates (i–iv), under the parameters: power = 0.4 mW, time = 30 s (i), a layer of silver nanoparticles with an average diameter of ~100 nm first formed on the ITO substrate. Then, for a longer irradiation time of ~60 s (ii), these nanoparticles became denser, and some nanorice structures began to emerge. Further increasing the irradiation time to 120 and 180 s (iii and iv) resulted in the nanorices growing larger and denser, with the length reaching up to ~400 nm. We then focused on the influence of irradiation power. The morphological changes that occurred as the laser power increased at certain irradiation times are shown by column. Along with aggregates i, v and ix, we also observed morphological changes from nanoparticles to nanorices. Based on these morphological evolutions, we deduced that the growth of silver aggregates was tuned by the photon dose through the combination effect of photoinduced growth and coalescence [53,69,70]: ① The silver nuclei were first formed in solution and then grown into nanoparticles through Ostwald ripening. ② As the nanoparticles grew, the adjacent particles began to coalesce with each other and formed linear polycrystalline structures called nanorices. Some concaves between the connected particles can still be observed from the SEM images of aggregates ii, iii, v and vi (see the magnified images in Figure S2 for details). ③ The nanorices grew via the atoms and/or nuclei addition. Finally, some nanoflake structures were formed. Here, we should note that the growth rate of the nanostructures can be tuned by the concentration of the reducing agent (citrate) [53]. As shown in Figure S3, we monitored the growth of silver aggregates under different citrate concentrations (0.01, 0.08, and 0.64 mM). The dark-field scattering images show that aggregates grew slowly under lower citrate concentrations. Under relatively high citrate concentrations, dense aggregates can form quickly in tens of seconds. In our experiments (data in Figure 3), a moderate citrate concentration (0.08 mM) was adopted to provide better controllability by the laser power and irradiation time.

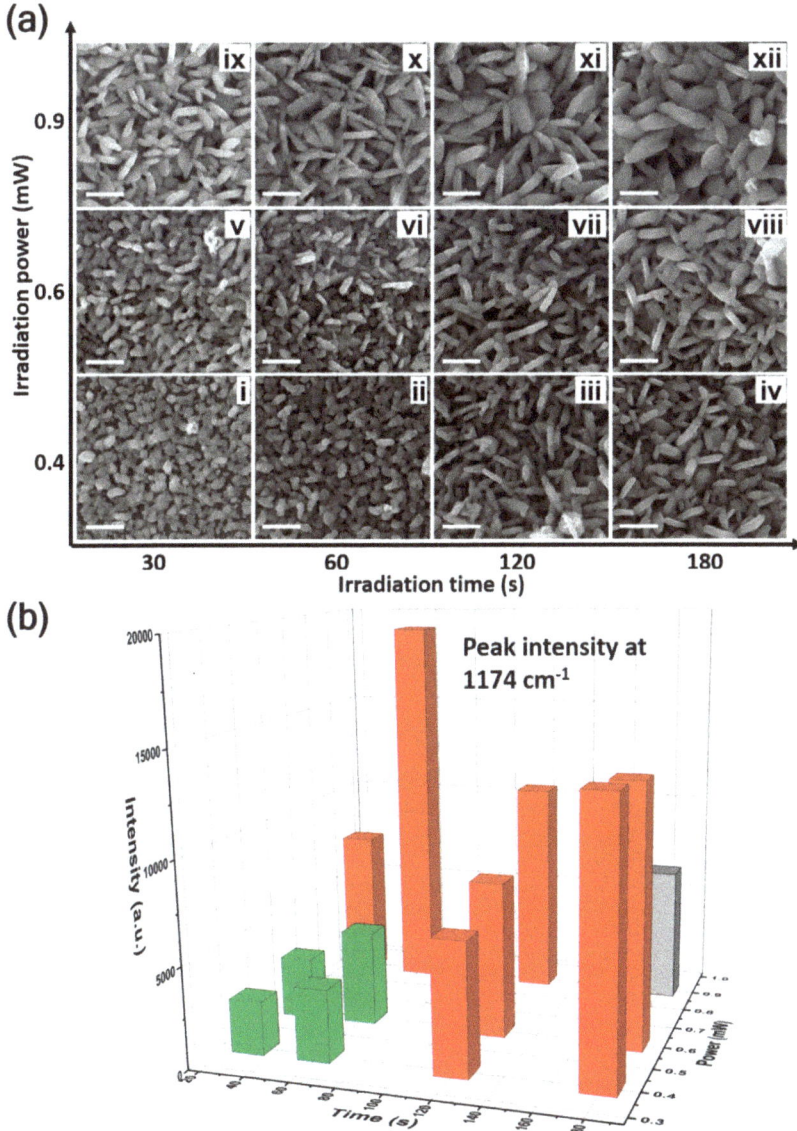

Figure 3. Controlling synthesis of silver aggregates and their SERS activities. (**a**) SEM images of silver aggregates grown under different irradiation power (0.4–0.9 mW) and exposure time (30–180 s). The scale bars are all 500 nm. (**b**) The corresponding SERS intensity from the silver aggregates are shown in (**a**).

Then, the SERS performance of these synthesized silver aggregates was investigated using 10^{-7} M CV as a probe. The corresponding SERS spectra are shown in Figure S4. Figure 3b presents the statistics of the peak intensity at 1174 cm^{-1}. Depending on the SERS intensity, the silver aggregates can be separated into two groups. For the first group (i, ii, v, vi), the enhancement was relatively low, with SERS intensity in the range of 2600–4600, where the aggregates were dominated by nanoparticles. Interestingly, the second group with compact nanorices aggregates (iii, iv, vii, viii, ix, x, xi) exhibited prominent SERS signals, among which the largest SERS intensity could reach up to 19,000 (x). From the SEM images in Figure 3a, we noticed that the aggregates x was composed of closely packed nanorice and flake structures with abundant surface roughness. This can be understood by the fact that the roughened nanorices and flakes aggregates can enhance the local field in two ways. One is the smaller nanogaps/crevices with higher near-field enhancement. On average, the compact nanostructures in group two (such as the aggregates x) generate smaller gaps than the sparsely small nanoparticles in group one (such as aggregates i), as shown in Figure S5. The other is from the abundant rough structures on the nanorices and flakes surface. The contribution of surface roughness to near-field enhancement on a mesostructure has also been confirmed by previous experiments and simulations [71–73]. Numerical simulations were also performed to help visualize the enhancement contributions from gaps and rough surface. As shown in Figure S6, the two coupled nanorices generated obvious near-field enhancement at both the gap and rough surface positions. While, the SERS intensity decreased in the case of aggregates xii, though the laser irradiation time (180 s) was longer than that of aggregate × (60 s). This could be caused by the disappearance of surface roughness during nanostructures overgrowth. The zoomed in images of aggregates x and xii are compared in Figure S5. Additionally, the overgrowth of nanostructures can also quench some nanogap enhancement due to the direct contact between the nanostructures.

4. Conclusions

In summary, a highly active silver aggregates SERS material was directly synthesized on the ITO–solution interface via a facile in situ photochemical reduction method. The morphologies of these aggregates were effectively controlled by laser power and irradiation time. By correlating the morphologies with their SERS signals, the best SERS aggregates were obtained under the synthesis parameters: power = 0.9 mW, time = 60 s. The average SERS EF was as large as 2.0×10^7. Importantly, the morphology features of optimal SERS aggregates were identified. Aggregates composed of packed nanorices and flakes with abundant surface roughness would possess better SERS ability. An understanding of the relation between morphology and SERS performance would be beneficial for controlled synthesis of optimal SERS materials with a high density of hot spots, and the development of practical SERS techniques.

Supplementary Materials: The following are available online at http://www.mdpi.com/2079-4991/9/11/1529/s1. Contents: SERS enhancement factor estimation. Figure S1. The absorption spectrum of 10^{-4} M CV in water; Figure S2. The morphological evolutions of silver aggregates; Figure S3. Monitoring the silver aggregates growths under different citrate concentrations; Figure S4. The SERS spectra measured on the silver aggregates synthesized under different parameters; Figure S5. The magnified images of aggregates i, x and xii; Figure S6. The numerical simulations of the local field distributions around the two coupled roughened nanorices.

Author Contributions: Z.L. and L.Y. supervised the experiments; J.Y.; X.C. and P.L. performed the experiments; Y.L. performed the numerical simulations; L.Y. and Z.L. analyzed the data; L.Y. and Z.L. wrote the manuscript. All authors discussed the results and commented on the manuscript.

Funding: This research was funded by the National Natural Science Foundation of China (Grant Nos. 11774245 and 11704266), the Fok Ying Tung Education Foundation, China (Grant No. 151010), the General Foundation of Beijing Municipal Commission of Education (Grant No. KM201810028006), the Beijing Natural Science Foundation (Z190006), the Training Program of the Major Research Plan of Capital Normal University, Yanjing Scholar Foundation of Capital Normal University, and the Scientific Research Base Development Program of Beijing Municipal Commission of Education.

Acknowledgments: This work was supported by the National Natural Science Foundation of China (Grant Nos. 11774245 and 11704266), the Fok Ying Tung Education Foundation, China (Grant No. 151010), the General Foundation of Beijing Municipal Commission of Education (Grant No. KM201810028006), the Beijing Natural Science Foundation (Z190006), the Training Program of the Major Research Plan of Capital Normal University, Yanjing Scholar Foundation of Capital Normal University, and the Scientific Research Base Development Program of Beijing Municipal Commission of Education.

Conflicts of Interest: The authors declare no conflict of interest.

References

1. Sharma, B.; Frontiera, R.R.; Henry, A.I.; Ringe, E.; Van Duyne, R.P. SERS: Materials, applications, and the future. *Mater. Today* **2012**, *15*, 16–25. [CrossRef]
2. Hakonen, A.; Andersson, P.O.; Schmidt, M.S.; Rindzevicius, T.; Käll, M. Explosive and chemical threat detection by surface-enhanced Raman scattering: A review. *Anal. Chim. Acta* **2015**, *893*, 1–13. [CrossRef] [PubMed]
3. Panneerselvam, R.; Liu, G.-K.; Wang, Y.-H.; Liu, J.-Y.; Ding, S.-Y.; Li, J.-F.; Wu, D.-Y.; Tian, Z.-Q. Surface-enhanced Raman spectroscopy: Bottlenecks and future directions. *Chem. Commun.* **2018**, *54*, 10–25. [CrossRef] [PubMed]
4. Lim, D.-K.; Jeon, K.-S.; Kim, H.M.; Nam, J.-M.; Suh, Y.D. Nanogap-engineerable Raman-active nanodumbbells for single-molecule detection. *Nat. Mater.* **2010**, *9*, 60–67. [CrossRef]
5. Xu, H.; Bjerneld, E.J.; Käll, M.; Börjesson, L. Spectroscopy of single hemoglobin molecules by surface-enhanced Raman scattering. *Phys. Rev. Lett.* **1999**, *83*, 4357. [CrossRef]
6. Gao, J.; Zhang, N.; Ji, D.; Song, H.; Liu, Y.; Zhou, L.; Sun, Z.; Jornet, J.M.; Thompson, A.C.; Collins, R.L.; et al. Superabsorbing metasurfaces with hybrid Ag–Au nanostructures for surface-enhanced Raman spectroscopy sensing of drugs and chemicals. *Small Methods* **2018**, *2*, 1800045. [CrossRef]
7. Yap, L.W.; Chen, H.; Gao, Y.; Petkovic, K.; Liang, Y.; Si, K.J.; Wang, H.; Tang, Z.; Zhu, Y.; Cheng, W. Bifunctional plasmonic-magnetic particles for an enhanced microfluidic SERS immunoassay. *Nanoscale* **2017**, *9*, 7822–7829. [CrossRef]
8. Cottat, M.; Lidgi-Guigui, N.; Tijunelyte, I.; Barbillon, G.; Hamouda, F.; Gogol, P.; Aassime, A.; Lourtioz, J.-M.; Bartenlian, B.; Chapelle, M.L.D.L. Soft UV nanoimprint lithography-designed highly sensitive substrates for SERS detection. *Nanoscale Res. Lett.* **2014**, *9*, 623. [CrossRef]
9. Sun, D.; Qi, G.; Xu, S.; Xu, W. Construction of highly sensitive surface-enhanced Raman scattering (SERS) nanosensor aimed for the testing of glucose in urine. *RSC Adv.* **2016**, *6*, 53800–53803. [CrossRef]
10. Lu, Y.; Zhou, T.; You, R.; Wu, Y.; Shen, H.; Feng, S.; Su, J. Fabrication and characterization of a highly-sensitive surface-enhanced Raman scattering nanosensor for detecting glucose in urine. *Nanomaterials* **2018**, *8*, 629. [CrossRef]
11. Zheng, P.; Li, M.; Jurevic, R.; Cushing, S.K.; Liu, Y.; Wu, N. A gold nanohole array based surface-enhanced Raman scattering biosensor for detection of silver (I) and mercury (II) in human saliva. *Nanoscale* **2015**, *7*, 11005–11012. [CrossRef] [PubMed]
12. Duan, J.; Yang, M.; Lai, Y.; Yuan, J.; Zhan, J. A colorimetric and surface-enhanced Raman scattering dual-signal sensor for Hg^{2+} based on bismuthiol II-capped gold nanoparticles. *Anal. Chim. Acta* **2012**, *723*, 88–93. [CrossRef] [PubMed]
13. Zhang, X.; Dai, Z.; Si, S.; Zhang, X.; Wu, W.; Deng, H.; Wang, F.; Xiao, X.; Jiang, C. Ultrasensitive SERS substrate integrated with uniform subnanometer scale "hot spots" created by a graphene spacer for the detection of mercury ions. *Small* **2017**, *13*, 1603347. [CrossRef]
14. Yang, H.; Ye, S.; Fu, Y.; Zhang, W.; Xie, F.; Gong, L.; Fang, P.; Chen, J.; Tong, Y. A simple and highly sensitive thymine sensor for mercury ion detection based on surface-enhanced Raman spectroscopy and the mechanism study. *Nanomaterials* **2017**, *7*, 192. [CrossRef] [PubMed]
15. Li, J.F.; Huang, Y.F.; Ding, Y.; Yang, Z.L.; Li, S.B.; Zhou, X.S.; Fan, F.R.; Zhang, W.; Zhou, Z.Y.; Wu, D.Y.; et al. Shell-isolated nanoparticle-enhanced Raman spectroscopy. *Nature* **2010**, *464*, 392–395. [CrossRef] [PubMed]
16. Tian, H.; Zhang, N.; Tong, L.; Zhang, J. In situ quantitative graphene-based surface-enhanced Raman spectroscopy. *Small Methods* **2017**, *1*, 1700126. [CrossRef]
17. Li, P.; Ma, B.; Yang, L.; Liu, J. Hybrid single nanoreactor for in situ sers monitoring of plasmon-driven and small Au nanoparticles catalyzed reactions. *Chem. Commun.* **2015**, *51*, 11394–11397. [CrossRef]
18. Sun, M.; Zhang, Z.; Zheng, H.; Xu, H. In-situ plasmon-driven chemical reactions revealed by high vacuum tip-enhanced Raman spectroscopy. *Sci. Rep.* **2012**, *2*, 647. [CrossRef]
19. Huang, W.; Jing, Q.; Du, Y.; Zhang, B.; Meng, X.; Sun, M.; Schanze, K.S.; Gao, H.; Xu, P. An in situ SERS study of substrate-dependent surface plasmon induced aromatic nitration. *J. Mater. Chem. C* **2015**, *3*, 5285–5291. [CrossRef]

20. Han, Q.; Zhang, C.; Gao, W.; Han, Z.; Liu, T.; Li, C.; Wang, Z.; He, E.; Zheng, H. Ag-Au alloy nanoparticles: Synthesis and in situ monitoring SERS of plasmonic catalysis. *Sens. Actuators B Chem.* **2016**, *231*, 609–614. [CrossRef]
21. Xu, H.; Aizpurua, J.; Käll, M.; Apell, P. Electromagnetic contributions to single-molecule sensitivity in surface-enhanced Raman scattering. *Phys. Rev. E* **2000**, *62*, 4318. [CrossRef] [PubMed]
22. McMahon, J.M.; Henry, A.-I.; Wustholz, K.L.; Natan, M.J.; Freeman, R.G.; Van Duyne, R.P.; Schatz, G.C. Gold nanoparticle dimer plasmonics: Finite element method calculations of the electromagnetic enhancement to surface-enhanced Raman spectroscopy. *Anal. Bioanal. Chem.* **2009**, *394*, 1819–1825. [CrossRef] [PubMed]
23. Li, Z.Y. Mesoscopic and microscopic strategies for engineering plasmon-enhanced Raman scattering. *Adv. Opt. Mater.* **2018**, *6*, 1701097. [CrossRef]
24. Ding, S.-Y.; You, E.-M.; Tian, Z.-Q.; Moskovits, M. Electromagnetic theories of surface-enhanced Raman spectroscopy. *Chem. Soc. Rev.* **2017**, *46*, 4042–4076. [CrossRef]
25. Ren, B.; Lin, X.-F.; Yang, Z.-L.; Liu, G.-K.; Aroca, R.F.; Mao, B.-W.; Tian, Z.-Q. Surface-enhanced Raman scattering in the ultraviolet spectral region: UV-SERS on rhodium and ruthenium electrodes. *J. Am. Chem. Soc.* **2003**, *125*, 9598–9599. [CrossRef]
26. Wu, D.-Y.; Liu, X.-M.; Duan, S.; Xu, X.; Ren, B.; Lin, S.-H.; Tian, Z.-Q. Chemical enhancement effects in SERS spectra: A quantum chemical study of pyridine interacting with copper, silver, gold and platinum metals. *J. Phys. Chem. C* **2008**, *112*, 4195–4204. [CrossRef]
27. Valley, N.; Greeneltch, N.; Van Duyne, R.P.; Schatz, G.C. A look at the origin and magnitude of the chemical contribution to the enhancement mechanism of surface-enhanced Raman spectroscopy (SERS): Theory and experiment. *J. Phys. Chem. Lett.* **2013**, *4*, 2599–2604. [CrossRef]
28. Halas, N.J.; Lal, S.; Chang, W.-S.; Link, S.; Nordlander, P. Plasmons in strongly coupled metallic nanostructures. *Chem. Rev.* **2011**, *111*, 3913–3961. [CrossRef]
29. Chen, S.; Meng, L.-Y.; Shan, H.-Y.; Li, J.-F.; Qian, L.; Williams, C.T.; Yang, Z.-L.; Tian, Z.-Q. How to light special hot spots in multiparticle–film configurations. *ACS Nano* **2015**, *10*, 581–587. [CrossRef]
30. Zhang, Y.-J.; Chen, S.; Radjenovic, P.; Bodappa, N.; Zhang, H.; Yang, Z.-L.; Tian, Z.-Q.; Li, J.-F. Probing the location of 3D hot spots in gold nanoparticle films using surface-enhanced Raman spectroscopy. *Anal. Chem.* **2019**, *91*, 5316–5322. [CrossRef]
31. Zong, S.; Chen, C.; Wang, Z.; Zhang, Y.; Cui, Y. Surface-enhanced Raman scattering based in situ hybridization strategy for telomere length assessment. *ACS Nano* **2016**, *10*, 2950–2959. [CrossRef] [PubMed]
32. Lu, H.; Zhu, L.; Zhang, C.; Chen, K.; Cui, Y. Mixing assisted "hot spots" occupying SERS strategy for highly sensitive in situ study. *Anal. Chem.* **2018**, *90*, 4535–4543. [CrossRef] [PubMed]
33. Liang, H.; Li, Z.; Wang, W.; Wu, Y.; Xu, H. Highly surface-roughened "flower-like" silver nanoparticles for extremely sensitive substrates of surface-enhanced Raman scattering. *Adv. Mater.* **2009**, *21*, 4614–4618. [CrossRef]
34. Liu, H.; Yang, Z.; Meng, L.; Sun, Y.; Wang, J.; Yang, L.; Liu, J.; Tian, Z. Three-dimensional and time-ordered surface-enhanced Raman scattering hot-spot matrix. *J. Am. Chem. Soc.* **2014**, *136*, 5332–5341. [CrossRef]
35. Liu, H.; Zhang, X.; Zhai, T.; Sander, T.; Chen, L.; Klar, P.J. Centimeter-scale-homogeneous SERS substrates with seven-order global enhancement through thermally controlled plasmonic nanostructures. *Nanoscale* **2014**, *6*, 5099–5105. [CrossRef]
36. Chen, Z.; Shi, H.; Wang, Y.; Yang, Y.; Liu, S.; Ye, H. Sharp convex gold grooves for fluorescence enhancement in micro/nano fluidic biosensing. *J. Mater. Chem. B* **2017**, *5*, 8839–8844. [CrossRef]
37. Barbillon, G. Fabrication and SERS performances of metal/Si and metal/ZnO nanosensors: A review. *Coatings* **2019**, *9*, 86. [CrossRef]
38. Zhao, X.; Deng, M.; Rao, G.; Yan, Y.; Wu, C.; Jiao, Y.; Deng, A.; Yan, C.; Huang, J.; Wu, S.; et al. High-performance SERS substrate based on hierarchical 3D Cu nanocrystals with efficient morphology control. *Small* **2018**, *14*, 1802477. [CrossRef]
39. Guo, Q.; Xu, M.; Yuan, Y.; Gu, R.; Yao, J. Self-assembled large-scale monolayer of Au nanoparticles at the air/water interface used as a SERS substrate. *Langmuir* **2016**, *32*, 4530–4537. [CrossRef]
40. Lee, Y.-J.; Schade, N.B.; Sun, L.; Fan, J.A.; Bae, D.R.; Mariscal, M.M.; Lee, G.; Capasso, F.; Sacanna, S.; Manoharan, V.N.; et al. Ultrasmooth, highly spherical monocrystalline gold particles for precision plasmonics. *ACS Nano* **2013**, *7*, 11064–11070. [CrossRef]

41. Li, X.; Zhang, J.; Xu, W.; Jia, H.; Wang, X.; Yang, B.; Zhao, B.; Li, B.; Ozaki, Y. Mercaptoacetic acid-capped silver nanoparticles colloid: Formation, morphology, and SERS activity. *Langmuir* **2003**, *19*, 4285–4290. [CrossRef]
42. Li, J.-M.; Yang, Y.; Qin, D. Hollow nanocubes made of Ag–Au alloys for SERS detection with sensitivity of 10^{-8} M for melamine. *J. Mater. Chem. C* **2014**, *2*, 9934–9940. [CrossRef]
43. Sun, Y.; Xia, Y. Shape-controlled synthesis of gold and silver nanoparticles. *Science* **2002**, *298*, 2176–2179. [CrossRef] [PubMed]
44. Kozuch, J.; Petrusch, N.; Gkogkou, D.; Gernert, U.; Weidinger, I.M. Calculating average surface enhancement factors of randomly nanostructured electrodes by a combination of SERS and impedance spectroscopy. *Phys. Chem. Chem. Phys.* **2015**, *17*, 21220–21225. [CrossRef]
45. Tao, A.; Sinsermsuksakul, P.; Yang, P. Polyhedral silver nanocrystals with distinct scattering signatures. *Angew. Chem. Int. Ed.* **2006**, *45*, 4597–4601. [CrossRef]
46. Liu, S.-Y.; Tian, X.-D.; Zhang, Y.; Li, J.-F. Quantitative surface-enhanced Raman spectroscopy through the interface-assisted self-assembly of three-dimensional silver nanorod substrates. *Anal. Chem.* **2018**, *90*, 7275–7282. [CrossRef]
47. Chen, Q.; Fu, Y.; Zhang, W.; Ye, S.; Zhang, H.; Xie, F.; Gong, L.; Wei, Z.; Jin, H.; Chen, J. Highly sensitive detection of glucose: A quantitative approach employing nanorods assembled plasmonic substrate. *Talanta* **2017**, *165*, 516–521. [CrossRef]
48. Kim, S.; Kim, D.-H.; Park, S.-G. Highly sensitive and on-site NO$_2$ SERS sensors operated under ambient conditions. *Analyst* **2018**, *143*, 3006–3010. [CrossRef]
49. Fang, Y.; Wei, H.; Hao, F.; Nordlander, P.; Xu, H. Remote-excitation surface-enhanced Raman scattering using propagating Ag nanowire plasmons. *Nano Lett.* **2009**, *9*, 2049–2053. [CrossRef]
50. Yang, S.; Dai, X.; Stogin, B.B.; Wong, T.-S. Ultrasensitive surface-enhanced Raman scattering detection in common fluids. *Proc. Natl. Acad. Sci. USA* **2016**, *113*, 268–273. [CrossRef]
51. Tang, S.; Li, Y.; Huang, H.; Li, P.; Guo, Z.; Luo, Q.; Wang, Z.; Chu, P.K.; Li, J.; Yu, X.-F. Efficient enrichment and self-assembly of hybrid nanoparticles into removable and magnetic SERS substrates for sensitive detection of environmental pollutants. *ACS Appl. Mater. Interfaces* **2017**, *9*, 7472–7480. [CrossRef] [PubMed]
52. Bjerneld, E.J.; Murty, K.; Prikulis, J.; Käll, M. Laser-induced growth of Ag nanoparticles from aqueous solutions. *Chem. Phys. Chem.* **2002**, *3*, 116–119. [CrossRef]
53. Bjerneld, E.J.; Svedberg, F.; Käll, M. Laser-induced growth and deposition of noble-metal nanoparticles for surface-enhanced Raman scattering. *Nano Lett.* **2003**, *3*, 593–596. [CrossRef]
54. Leopold, N.; Lendl, B. On-column silver substrate synthesis and surface-enhanced Raman detection in capillary electrophoresis. *Anal. Bioanal. Chem.* **2010**, *396*, 2341–2348. [CrossRef]
55. Herman, K.; Szabó, L.; Leopold, L.F.; Chiş, V.; Leopold, N. In situ laser-induced photochemical silver substrate synthesis and sequential SERS detection in a flow cell. *Anal. Bioanal. Chem.* **2011**, *400*, 815–820. [CrossRef]
56. Xu, B.-B.; Ma, Z.-C.; Wang, L.; Zhang, R.; Niu, L.-G.; Yang, Z.; Zhang, Y.-L.; Zheng, W.-H.; Zhao, B.; Xu, Y.; et al. Localized flexible integration of high-efficiency surface-enhanced Raman scattering (SERS) monitors into microfluidic channels. *Lab Chip* **2011**, *11*, 3347–3351. [CrossRef]
57. Xie, Y.; Yang, S.; Mao, Z.; Li, P.; Zhao, C.; Cohick, Z.; Huang, P.-H.; Huang, T.J. In situ fabrication of 3D Ag@ZnO nanostructures for microfluidic surface-enhanced Raman scattering systems. *ACS Nano* **2014**, *8*, 12175–12184. [CrossRef]
58. Ma, Z.-C.; Zhang, Y.-L.; Han, B.; Liu, X.-Q.; Zhang, H.-Z.; Chen, Q.-D.; Sun, H.-B. Femtosecond laser direct writing of plasmonic Ag/Pd alloy nanostructures enables flexible integration of robust SERS substrates. *Adv. Mater. Technol.* **2017**, *2*, 1600270. [CrossRef]
59. Yan, W.; Yang, L.; Chen, J.; Wu, Y.; Wang, P.; Li, Z. In situ two-step photoreduced SERS materials for on-chip single-molecule spectroscopy with high reproducibility. *Adv. Mater.* **2017**, *29*, 1702893. [CrossRef]
60. Lee, G.P.; Bignell, L.J.; Romeo, T.C.; Razal, J.M.; Shepherd, R.L.; Chen, J.; Minett, A.I.; Innis, P.C.; Wallace, G.G. The citrate-mediated shape evolution of transforming photomorphic silver nanoparticles. *Chem. Commun.* **2010**, *46*, 7807–7809. [CrossRef]
61. Condorelli, M.; Scardaci, V.; D'Urso, L.; Puglisi, O.; Fazio, E.; Compagnini, G. Plasmon sensing and enhancement of laser prepared silver colloidal nanoplates. *Appl. Surf. Sci.* **2019**, *475*, 633–638. [CrossRef]
62. Elechiguerra, J.L.; Reyes-Gasga, J.; Yacaman, M.J. The role of twinning in shape evolution of anisotropic noble metal nanostructures. *J. Mater. Chem.* **2006**, *16*, 3906–3919. [CrossRef]

63. Cañamares, M.V.; Chenal, C.; Birke, R.L.; Lombardi, J.R. DFT, SERS, and single-molecule SERS of crystal violet. *J. Phys. Chem. C* **2008**, *112*, 20295–20300. [CrossRef]
64. Xing, G.; Wang, K.; Li, P.; Wang, W.; Chen, T. 3D hierarchical Ag nanostructures formed on poly (acrylic acid) brushes grafted graphene oxide as promising SERS substrates. *Nanotechnology* **2018**, *29*, 115503. [CrossRef] [PubMed]
65. Zhang, X.; Xiao, X.; Dai, Z.; Wu, W.; Zhang, X.; Fu, L.; Jiang, C. Ultrasensitive SERS performance in 3D "sunflower-like" nanoarrays decorated with Ag nanoparticles. *Nanoscale* **2017**, *9*, 3114–3120. [CrossRef] [PubMed]
66. Bryche, J.-F.; Bélier, B.; Bartenlian, B.; Barbillon, G. Low-cost SERS substrates composed of hybrid nanoskittles for a highly sensitive sensing of chemical molecules. *Sens. Actuators B Chem.* **2017**, *239*, 795–799. [CrossRef]
67. Wiley, B.J.; Chen, Y.; Mclellan, J.M.; Xiong, Y.; Li, Z.; Ginger, D.; Xia, Y. Synthesis and optical properties of silver nanobars and nanorice. *Nano Lett.* **2007**, *7*, 1032–1036. [CrossRef] [PubMed]
68. Xu, H.; Käll, M. Polarization-dependent surface-enhanced Raman spectroscopy of isolated silver nanoaggregates. *Chem. Phys. Chem.* **2003**, *4*, 1001–1005. [CrossRef]
69. Pong, B.-K.; Elim, H.I.; Chong, J.-X.; Ji, W.; Trout, B.L.; Lee, J.-Y. New insights on the nanoparticle growth mechanism in the citrate reduction of gold (III) salt: Formation of the Au nanowire intermediate and its nonlinear optical properties. *J. Phys. Chem. C* **2007**, *111*, 6281–6287. [CrossRef]
70. Liu, T.; Xiao, X.; Yang, C. Surfactantless photochemical deposition of gold nanoparticles on an optical fiber core for surface-enhanced Raman scattering. *Langmuir* **2011**, *27*, 4623–4626. [CrossRef]
71. Wang, H.; Halas, N.J. Mesoscopic Au "meatball" particles. *Adv. Mater.* **2008**, *20*, 820–825. [CrossRef]
72. Fang, J.; Du, S.; Lebedkin, S.; Li, Z.; Kruk, R.; Kappes, M.; Hahn, H. Gold mesostructures with tailored surface topography and their self-assembly arrays for surface-enhanced Raman spectroscopy. *Nano Lett.* **2010**, *10*, 5006–5013. [CrossRef] [PubMed]
73. Liu, Z.; Zhang, F.; Yang, Z.; You, H.; Tian, C.; Li, Z.; Fang, J. Gold mesoparticles with precisely controlled surface topographies for single-particle surface-enhanced Raman spectroscopy. *J. Mater. Chem. C* **2013**, *1*, 5567–5576. [CrossRef]

© 2019 by the authors. Licensee MDPI, Basel, Switzerland. This article is an open access article distributed under the terms and conditions of the Creative Commons Attribution (CC BY) license (http://creativecommons.org/licenses/by/4.0/).

Article

Surface-Enhanced Raman Scattering and Fluorescence on Gold Nanogratings

Yu-Chung Chang *, Bo-Han Huang and Tsung-Hsien Lin

Department of Electrical Engineering, National Changhua University of Education, Changhua 500, Taiwan
* Correspondence: ycchang@cc.ncue.edu.tw

Received: 23 March 2020; Accepted: 15 April 2020; Published: 17 April 2020

Abstract: Surface-enhanced Raman scattering (SERS) spectroscopy is a sensitive sensing technique. It is desirable to have an easy method to produce SERS-active substrate with reproducible and robust signals. We propose a simple method to fabricate SERS-active substrates with high structural homogeneity and signal reproducibility using electron beam (E-beam) lithography without the problematic photoresist (PR) lift-off process. The substrate was fabricated by using E-beam to define nanograting patterns on the photoresist and subsequently coat a layer of gold thin film on top of it. Efficient and stable SERS signals were observed on the substrates. In order to investigate the enhancement mechanism, we compared the signals from this substrate with those with photoresist lifted-off, which are essentially discontinuous gold stripes. While both structures showed significant grating-period-dependent fluorescence enhancement, no SERS signal was observed on the photoresist lifted-off gratings. Only transverse magnetic (TM)-polarized excitation exhibited strong enhancement, which revealed its plasmonic attribution. The fluorescence enhancement showed distinct periodic dependence for the two structures, which is due to the different enhancement mechanism. We demonstrate using this substrate for specific protein binding detection. Similar periodicity dependence was observed. Detailed theoretical and experimental studies were performed to investigate the observed phenomena. We conclude that the excitation of surface plasmon polaritons on the continuous gold thin film is essential for the stable and efficient SERS effects.

Keywords: surface-enhanced Raman scattering (SERS); localized surface plasmon resonance (LSPR); surface plasmon polariton (SPP); surface plasmon resonance (SPR); nanograting; nanofabrication; electron beam lithography

1. Introduction

Surface-enhanced Raman scattering (SERS) spectroscopy is a powerful analytic tool for sensitive molecular quantification. Since its discovery by Fleischmann et al. on a roughened silver surface in 1974 [1], it has attracted a lot of attention because the significantly enhanced Raman signal is very helpful for specific identification of chemical and biological molecules [2]. The greatly enhanced signal is due to the interaction between the incident light and the nanometer-sized metallic structure, which gives rise to a significant enhancement in the local field at the metal surfaces due to the excitation of surface plasmon resonance (SPR) [3]. An enormous enhancement factor (EF) up to 10^{14} was achieved and single molecule detection was demonstrated independently by Nie and Kneipp et al. in 1997 using SERS [4,5]. The research activity in the field of plasmonic enhanced spectroscopy was boosted since then in the last two decades. The high sensitivity of SERS has been exploited in many applications, such as chemistry, physical and biological sciences, environmental monitoring, and medical diagnostics, etc. [2,6,7]. Because of the promising potential of SERS-based sensing, we have witnessed exponentially increased research activities in this field [8]. For SERS-sensing to have real-world impact, it is essential to have reproducible, large-area, and cost-effective SERS-active substrates.

The keys to enhance the electromagnetic near field for SERS spectroscopy are nanoscaled surface roughness and nanostructures. Various methods have been developed to fabricate SERS substrates. The fabrication techniques can be classified into two categories. The bottom-up techniques are typically based on chemical synthesis. They are usually cost-efficient and easily accessible. For example, noble metal nanoparticle (NP) self-assembly is capable of producing highly ordered NP with various shapes for efficient SERS detection [9,10]. However, these chemical methods usually have an uncontrolled NP aggregation process, which results in poor reproducibility of the SERS signal. The substrate inhomogeneity significantly limits their applications [11]. On the other hand, the nanofabrication-based, top-down techniques, such as optical [12] and electron beam (E-beam) lithography (EBL) [13,14], focused ion beam [15], and nanoimprint lithography [15,16], can produce substrates with high structural homogeneity and SERS signal reproducibility that are most desired for modern SERS applications [17,18].

Although reproducible SERS substrates with high EF could be achieved with these sophisticated nanofabrication techniques, a lift-off process of the photoresist is often necessary to produce the nanoscale features. This process is time-consuming and challenging for large-area production. For most SERS-active substrates, the enhancement is based on localized surface plasmon resonance (LSPR) on the nanoscale features, where the electrical field is greatly enhanced [19,20]. Usually, the smaller the feature the better the enhancement is when the structural geometry is optimized. However, the failure rate of the lift-off process for such small features is much higher, which thus renders a low yield of production. For commercial applications, it is desirable to have a SERS-active substrate, which is easy to fabricate, cost-effective, and reproducible with high yield.

Here, we present a facile method to produce a reliable SERS-active substrate without the time-consuming and troublesome lift-off process. The obtained SERS signal is reproducible and robust. We use an E-beam to define highly ordered nanograting patterns on the photoresist and coated the surface with a thin layer of gold. We demonstrate highly efficient SERS signals from the substrates. The grating structure enables the excitation of surface plasmon polaritons (SPPs) to propagate on the corrugated metal surface when the Bragg condition is satisfied [21].

In order to understand the origin of the enhancement effects, we compared the signals from the above-mentioned SERS substrate with photoresist (PR) lifted-off gold nanogratings, which is essentially discontinuous gold nanostripes. SERS signal is only observed on the gratings without PR lifted-off. We observed enhanced fluorescence signals for both structures, but they exhibited distinct periodicity dependence. This implied that the enhancements might be due to dissimilar mechanisms. Efficient enhancements were only observed for transverse magnetic (TM)-polarized excitation, which revealed its plasmonic origin. We systematical investigated the periodicity dependence of the enhanced Raman scattering and fluorescence signals. The signal strength of SERS and fluorescence enhancement showed similar periodicity dependence on the substrate without PR lifted-off, which indicated that they originated from the same enhancement effect. Our simulation and theoretical treatments agreed well with the experimental results. We found that the excitation of surface plasmon polariton on the continuous gold thin film was essential for the stable and efficient SERS effects, while the fluorescence on the lifted-off substrates was due to localized surface plasmon resonance. This facile method can be readily employed to produce cost-effective, large-area, SERS-active substrates with high throughput and reproducibility for practical applications.

2. Materials and Methods

The nanograting substrates were fabricated by the following procedures as illustrated in Figure 1. A layer of 200 nm-thick polymethylmethacrylate (PMMA) (Microchem A4, Kayaku Advanced Materials, Westborough, MA, USA) photoresist was spin-coated onto the ITO (indium tin oxide) glass. After prebaking to 180 °C to remove moisture and residue chemicals, we used E-beam lithography to define the nanograting patterns on the photoresist with various periods from 100 nm to 800 nm. A duty cycle of 50% was chosen for all periodicities because it had the best enhancement efficiency [22]. After the development process, we obtained periodic PMMA nanostripes as shown in Figure 2a. The grating size was 100 × 100 µm² for each periodicity. In this study, we investigated SERS and fluorescence signals from two kinds of nanograting structures as shown in Figure 1: **A**. nonlift-off and **B**. lift-off. For structure **A**, the developed sample was sputtered with a layer of gold thin film on top of it without a lift-off process. The scanning electron microscope (SEM) image of a gold thin film coated 200 nm period substrate is shown in Figure 2b. For structure **B**, the photoresist was removed by a typical lift-off process, leaving periodic gold nanostripes, which were separated from each other. The thickness of the gold thin film was 56 nm for all samples as it gave the best SERS efficiency for our structure. The samples were subsequently coated with a thin layer (14 nm) of PMMA (Microchem A9) as the spacer layer to reduce the quenching effect.

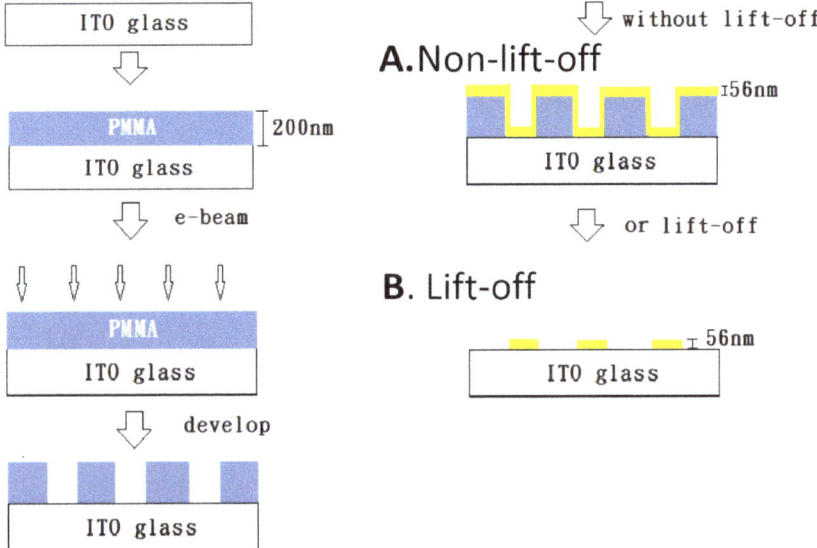

Figure 1. Schematic of the sample fabrication procedure. A layer of polymethylmethacrylate (PMMA) photoresist is first spin coated on the ITO glass. Nanograting patterns of the PMMA photoresist are fabricated by E-beam lithography. A layer of 56 nm gold thin film is subsequently sputtered on the nanostructure to obtain the structure **A** (nonlift-off). If the photoresist is removed by a lift-off process, we obtain the structure **B** (lift-off), which is discrete gold nanostripes.

For the spectroscopic studies, Rhodamin-6G (R6G) molecules (Sigma-Aldrich, St. Louis, MO, USA) were mixed in Milli-Q water (Merck, Darmstadt, Germany) to make a 10 µM solution and spin-coated on the substrate surface for subsequent measurements. Spectroscopic measurements were conducted with a confocal Raman microscope (DXR Raman, Thermo Fisher Scientific, Waltham, MA, USA). The excitation wavelength was 532 nm. The laser power was 1 mW unless mentioned otherwise. The excitation laser is linearly polarized. We rotate the samples to change the direction of excitation. For the protein-binding experiment, we add a quarter-wave plate in the optical path to make the excitation polarization circular to avoid the influence of sample orientation. The signals were collected with a 0.5NA 50× long working distance lens. The integration time was 1 s. The short measurement time prevented drying and heating of the sample. For each experimental condition, the spectra were measured at least 10 times.

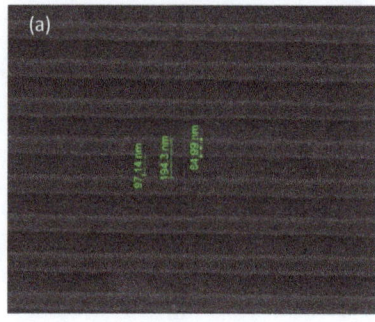 

Figure 2. Scanning electron microscope (SEM) images of the fabricated samples. (**a**) 200 nm-period PMMA nanograting. (**b**) Gold thin film coated 200 nm-period nanograting.

In order to investigate the ability to use the substrate for biologically relevant specific binding sensing, we deposited a layer of biotinylated bovine serum albumin (BSA) on top of the gold nanogratings by immersing the sample in a 10 mg/mL BSA solution (Sigma-Aldrich, St. Louis, MO, USA) for 2 h. The BSA is sticky and easily binds to the gold surface with a layer thickness of about 3–4 nm. After raising with Milli-Q water, we immersed the samples in a 10 µM Rhodamine-conjugated streptavidin phosphate buffered saline (PBS) solution (Invitrogen, Thermo Fisher Scientific, Waltham, Massachusetts, USA) for a period of 30 min. The labeled proteins could homogenously bind to the surface of the substrates through the specific interaction between biotin and streptavidin. The total protein layer thickness was about 7–8 nm [23], which acted as the spacer layer in this case. Finally we rinsed away unbound proteins. During the microscopic measurements, the samples were maintained wet by immersing it in a thin layer of PBS solution.

In order to understand the enhancement mechanism and to verify the experimental results, we used COMSOL Multiphysics (COMSOL Inc., Stockholm, Sweden) to simulate the two kinds of nanograting structures. This numerical calculation was based on the finite element method (FEM). The calculations were performed on a single unit cell. Because the size of the grating (100 µm) was much larger than the grating period, the longitudinal length of the nanostripes could be assumed as infinitely long compared to the grating period. Hence we simplified the problem into a 2D geometry to significantly relax the required computing power and greatly reduce the calculation time. A TM-polarized plane wave (magnetic field parallel to the nanostripes) was set as the excitation source. As for material parameters used in the simulation, the complex relative permittivity of gold thin film was interpolated from ref. [24] as $\varepsilon_{Au} = -5.33 + 2.55i$ at 532 nm. The refractive index of PMMA, ITO and water used in the simulation were 1.49, 1.88 and 1.33 respectively. In order to avoid singularities of simulation and to better model our structure, all corners are rounded by fillets in COMSOL. The corners of the PR gratings were rounded with a radius of curvature of 20 nm. The corners of gold gratings had a radius of curvature of 25 nm. For gold thin film coated PR gratings, the thickness of the gold thin film on the side walls of the PR is 15 nm, while the thickness of the gold thin film on top and in the grooves of the PR gratings was 56 nm as specified previously.

3. Results and Discussion

The measured spectra from nanogratings with various periodicities are shown in Figure 3. Figure 3a shows spectra from the nonlift-off substrates (structure A) and Figure 3b shows spectra from the lift-off substrates (structure B). Only nonlift-off nanogratings gave prominent Raman signals. As clearly seen in the figures, the signal strength is highly dependent on the grating period. For both cases, the highest enhancement on the fluorescence signals was about one order of magnitude compared to the case without corrugated grating structure (no-grating). The fluorescence signals on the nonlift-off gratings were about half of that of the lift-off samples, which should be due to fluorescence quenching as the gold coated area of the lift-off samples was exactly half of that of the nonlift-off ones. For the nonlift-off samples, the no-grating case was gold thin film coated flat PMMA, while for the lift-off samples, the no-grating case was bare ITO glass. By carefully analyzing the fluorescence signals, we noticed an obvious redshift of the spectra peak on the enhanced fluorescence. For both structures, the higher the enhancement, the more the redshift was, as was often reported in the literature of surface plasmon-enhanced fluorescence [25]. Therefore, the observed strong Raman signal could be attributed to surface plasmon-enhanced scattering (SERS).

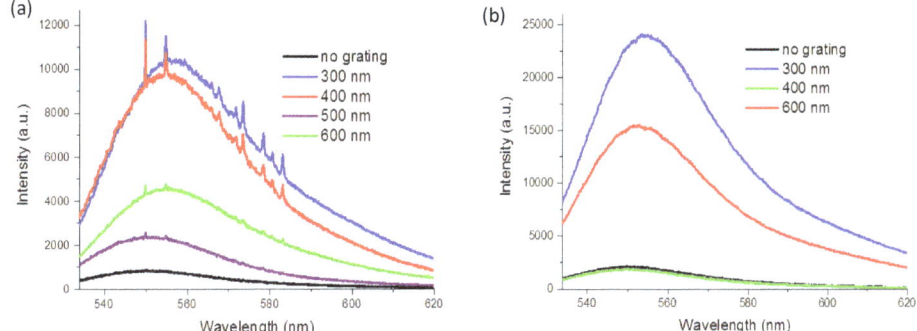

Figure 3. Raw data from (**a**) nonlift-off and (**b**) lift-off nanogratings of representative periods. The black curves are the spectra from substrates with no corrugated grating structure. Surface-enhanced Raman scattering (SERS) signal is only observable on the nonlift-off nanogratings. Prominent periodicity dependence of the Raman and fluorescence signals is manifested.

When we compare the fluorescence signals from the two kinds of nanogratings, we found very interesting phenomena. Their fluorescence enhancement had distinct periodicity dependence. For example, although both had the highest enhancement at a period of 300 nm, for 400 nm period samples, the nonlift-off grating had a similar enhancement as the 300 nm, but the signal from the lift-off grating was even smaller than the one without a grating (control). The fluorescence intensity for the 600 nm-period lift-off grating increased again while the signal from 600 nm-period nonlift-off gratings decreased. The drastic difference intrigued us to investigate the enhancement mechanism of the two structures by the fluorescence signals. Because Rhodamine-6G (R6G) has a strong Raman signal and high quantum efficiency, we chose it and excited at 532 nm to simultaneously observe the evolution of SERS and fluorescence signals at different substrate conditions. For solely SERS applications or studies, a 785 nm laser excitation could be employed to avoid the problem of the fluorescence background. It was noted that the fluorescence intensity typically reduced rapidly in time over a few seconds immediately after the excitation radiation was switched on. This was because of the accumulation of electrons in nonradiative triplet states of the fluorophore [12,26]. However, the Raman signals were quite stable over time. Therefore, although R6G had a strong fluorescence background, it did not affect our SERS measurements.

For our measurement conditions, no Raman signal could be found on lift-off gratings. Therefore, the following discussions about SERS are specifically for nonlift-off gratings. Figure 4a shows background subtracted SERS signals from nonlift-off nanogratings. The Raman features nicely correspond to typical R6G Raman signals [21]. We noticed that sometimes there were small Raman signals even without the corrugated grating structure. This might be due to the roughness of the PR surface. The gold thin film on flat glass substrate was checked by atomic force microscopy to have a smoothness better than 1/1000. The irregular nanoscale features might have induced enhanced local fields due to localized surface plasmon resonance [27]. However, for a short integration time of 1 s, the Raman signal on gratingless samples was often not noticeable. As will be discussed below, we attributed the greatly enhanced Raman signal to the coupling of LSPR and SPP, where SPP plays an important role in the SERS effect on the nonlift-off grating [11,28]. With the help of grating, SPP can be excited on the gold surface to propagate on the metal–dielectric interface. Figure 4b summarizes the dependence of Raman signal strength versus periodicity and excitation polarization. We noticed strong polarization dependence on the Raman signals. As is evident from the figure, the Raman signals are only moderately enhanced for TE (transverse electric)-polarized excitation and there is no obvious periodical dependence for them. On the contrary, for TM-polarized excitation (magnetic field parallel to the grating grooves), we have more than one order of magnitude enhancement for gratings of periods 200, 300 and 400 nm. Conspicuous periodicity dependence can be identified in the figure. The periodicity dependence is similar among different sets of samples. The SERS signal is reproducible and robust for samples of the same conditions. The standard deviation is less than 20% of the mean value for all cases.

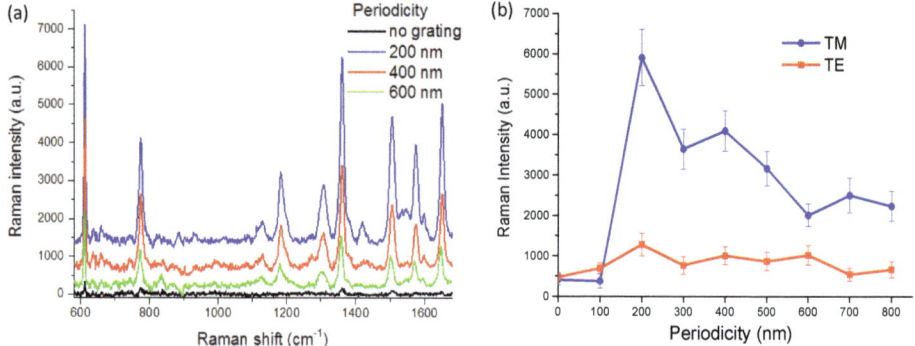

Figure 4. (**a**) Background corrected Raman signal from nonlift-off nanogratings. The spectra are labeled with different color respectively. The black line is the spectrum from a gratingless substrate, which is gold thin film coated flat photoresist (PR) layer (no-grating). The blue line is the spectrum from a 200 nm-period grating. Red line is the spectrum from a 400 nm-period grating. Green line is the spectrum from a 600 nm-period grating. The spectra are vertically shifted for visual clarity. (**b**) The periodicity dependence of Raman intensity for transverse magnetic (TM)- (blue circles) and transverse electric (TE)-polarized (red squares) excitation.

As mentioned previously, the different periodicity dependence of the fluorescence enhancement for the two kinds of nanograting structures is intriguing for further investigation. In Figure 5, we compare the dependence of the fluorescence signal intensity on the excitation light polarization and grating periodicity for the two kinds of nanogratings. We see a drastic difference between the two kinds of structures, which implies the enhancements might be due to distinct plasmonic effects. For both cases, only TM-polarized excitation exhibited strong enhancement at certain periodicities. The maximum enhancements compared to the gratingless substrate were both on the order of 10 times. The nonlift-off gratings had a higher enhancement factor compared to the lift-off gratings. However, the magnitude of fluorescence intensity on the nonlift-off grating was smaller than that on the lift-off grating. This was due to a higher quenching rate from the continuous gold thin film on the nonlift-off gratings, while the lift-off gratings were only 50% covered. As seen in the figure, the TE-polarized excitation also moderately increased the signal. This is because when the phase matching condition of Wood's anomaly is satisfied, the grating couples light onto the grating [29], which increases the excitation efficiency. However, only TM-polarized excitation can excite SPP on the gold–dielectric intersurface. The excited surface plasmon can concentrate the optical field at the vicinity of the gold surface, which results in a greatly magnified electrical field. For the nonlift-off grating, the fluorescence signal was significantly enhanced for periods 200, 300, and 400 nm. As the period increased, the signal decreased and reached a minimum when the period was 500 nm. At periods of 600 nm and 700 nm, the fluorescence intensities increased again, but not as high as in the shorter period cases. This periodicity dependence is due to matching the dispersion relation of grating coupled SPP excitation as will be discussed later. However, for the lift-off gratings, as seen in Figure 5b, the signal maximum occurred at 300 and 600 nm periods. The fluorescence intensity varies with periodicity acutely. We attribute the enhancement in the case of lift-off gratings to localized surface plasmon resonances (LSPR) because the enhancement only occurred when the rigorous resonance condition was satisfied. The dispersion relation for LSPR is

$$k_{sp} = \frac{2\pi}{\lambda_{sp}} = \frac{2\pi}{\lambda_0} \sqrt{\frac{\varepsilon_{Au}\varepsilon_d}{\varepsilon_{Au} + \varepsilon_d}} \quad (1)$$

where k_{sp} is the surface plasmon wave vector, λ_{sp} is the surface plasmon wavelength, λ_0 is the free space light wavelength, ε_{Au} is the dielectric function of gold, and ε_d is the dielectric function of the surrounding dielectric medium. For the excitation wavelength (λ_0) of 532 nm, the resonant surface plasmon wavelength at the PMMA-gold interface is 288.3 nm.

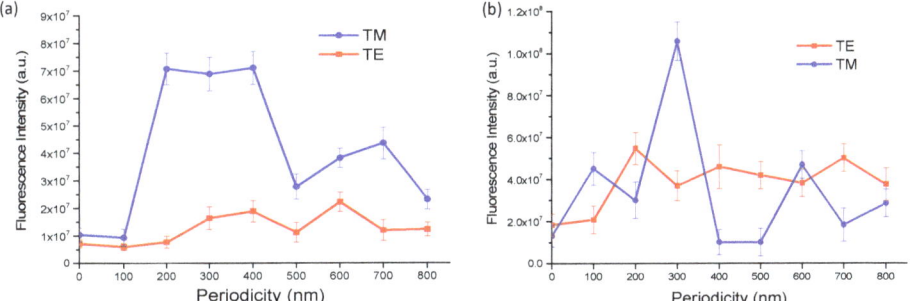

Figure 5. Periodicity and polarization dependence of the fluorescence intensities for (**a**) nonlift-off and (**b**) lift-off nanogratings. The blue circles and lines are for TM-polarized excitation and the red squares and lines are for TE-polarized excitation.

When the surface plasmon wavelength matches the transverse dimension of the gold nanostripe, it forms SP standing waves on the metal-dielectric interface [23]. The condition for a resonant standing SP wave across a nanostripe is

$$k_{sp} \cdot \frac{\Lambda}{2} = m\pi \quad (2)$$

where Λ is the grating period and m is an integer. Therefore, for the lift-off grating, the first peak at 300 nm corresponds to the lowest order mode when m = 1, and the peak at 600 nm corresponds to the m = 2 order. The different excitation mechanisms might account for the drastically different periodicity dependence between the two structures.

When comparing the enhancement of Raman and fluorescence signals of the nonlift-off substrate, Figures 4b and 5a, the periodicity dependence almost coincide with each other. They both had significant enhancement in a broad range of grating period from 200–400 nm, and another peak at about 700 nm. Therefore, we might infer that they were enhanced due to the same mechanism. For Raman signals, it seems the signal was larger for smaller periods and signal at the period of 200 nm was significantly enhanced. This might be due to a combined effect of SPP and LSPR coupling since the electromagnetic fields around small features are greatly enhanced by localized surface plasmons [30]. This will be discussed together with the simulation results later.

Our proposed method is capable of producing homogeneous, SERS-active substrates with reproducible and stable signals that are suitable for biological sensing. We prepared our substrate for the specific sensing of proteins using the protocol described in the Materials and Methods section. Here we used circularly polarized light for excitation to ease the sample preparation and orientation restrictions. For proof-of-principle, we exploited the specific protein binding system of biotin and streptavidin [31]. The gold nanograting surface was first deposited with a monolayer of biotinylated BSA and subsequently immersed in Rhodamine-tagged streptavidin for the interaction to take place.

The measured Raman signal is shown in Figure 6a. An intense SERS signal was only observed on the nonlift-off nanogratings and the enhancement was highly dependent on the periodicity, as expected. The maximum SERS occurred at a period of 300 nm. The maximum enhancement ratio for Raman was similar to the substrates described above. However, the intensities of Raman and fluorescence signals were both smaller. This is due to quenching caused by the gold thin film. There was no 14 nm-thick PMMA spacer layer above the gold thin film. The self-adsorbed protein layer was acting as the spacer layer. Although the fluorescence intensity was smaller, the enhancement factor for SERS was the same. The signal was adequate for molecular quantification. The use of proteins as the spacer layer simplified the SERS-active substrate fabrication and was convenient for typical biochemical laboratories.

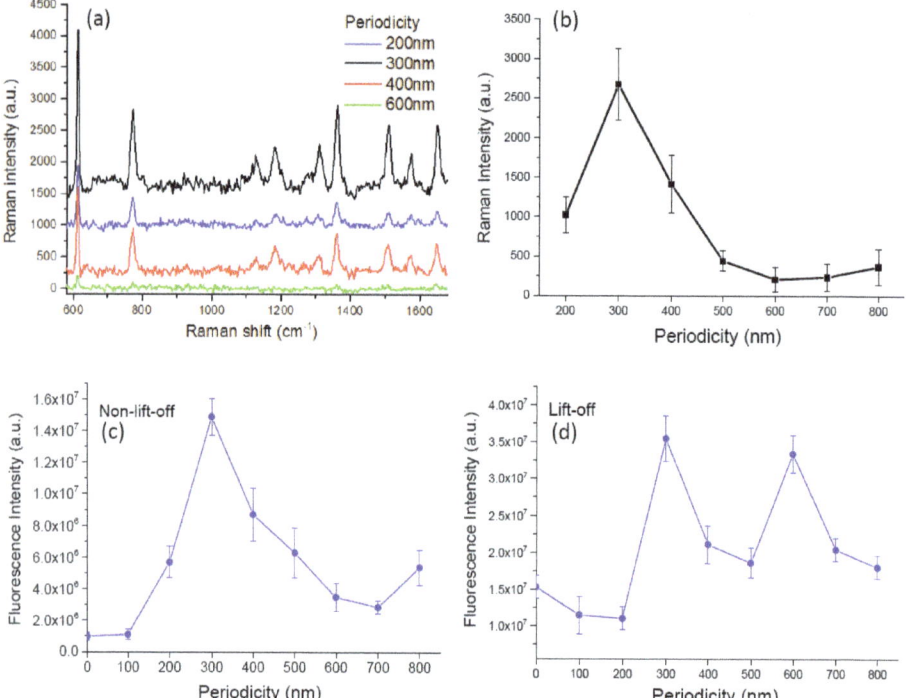

Figure 6. (**a**) Background corrected SERS spectra from nonlift-off nanogratings labeled with proteins. The spectra are vertically shifted for visual clarity. (**b**) The dependence of Raman intensity versus grating periodicity. (**c**) The periodicity dependence of fluorescence intensity on nonlift-off nanogratings. (**d**) The periodicity dependence of fluorescence intensity on lift-off nanogratings.

Here, we found the optimal periodicity for Raman and fluorescence was both 300 nm for nonlift gratings, which was different from the case with a PMMA spacer. Besides, in the case with a PMMA spacer on top of the gold surface, we had a second peak at a period of 700 nm. Here we had a minimum at 700 nm for both Raman and fluorescence and the second maximum seemed to occur at a longer period. This could be reasonably explained by the different SP wavelengths on the water–gold and PMMA–gold interfaces. Because there was no PMMA spacer layer, the medium above the gold thin film was water. The SP wavelength for the water–gold interface can be calculated by the dispersion relation of Equation (1). The SPP wavelength at a water–gold interface is 337.6 nm, which is longer than the PMMA–gold interface of 288.3 nm. Therefore the resonance conditions shifted to a longer period. Figure 6c,d are the periodicity dependence of fluorescence intensities for nonlift-off and lift-off nanogratings, respectively. Similarly, we see a distinct periodicity dependence between them. For the two kinds of structures, they respectively exhibit similar periodicity dependence as the case with PMMA spacer. For the nonlift-off case, it seems the resonance condition could be satisfied for a broader range of period from 200–400 nm. The enhancement decreases as the period becomes longer and increases again at about 800 nm. The same trends are seen in Raman and fluorescence signals. On the other hand, for lift-off samples, we see two peaks at periods of 300 and 600 nm. The trends are again similar to the case with a PMMA spacer on top of the grating. However, for the current condition, the 600 nm period grating exhibited almost the same enhancement as the 300 nm period grating. It seems a better resonance condition was achieved. This is reasonable if we consider the longer SP wavelength of the water–gold interface. The actual length of the interface was longer than half of the period because the gold layer had a thickness of 56 nm and the corners were usually rounded. The longer SP wavelength better fitted the resonant standing wave condition. Because of the strict resonance condition, the LSPR induced enhancement only occurred at specific periods.

In the above studies, we observed more than one order of magnitude enhancement of Raman and fluorescence intensities on nonlift nanogratings, while the maximum fluorescence enhancement on lift-off nanogratings was only about fivefold. To have a deeper understanding of the experimental results, we conducted numerical simulation on the two kinds of nanograting structures using the finite element method. The simulation results are shown in Figure 7. The simulated model was the structure without the PMMA spacer layer on top of the gold layer to simplify the problem and make the discussion concise. The medium above the gold layer was water. The plots show the average squared electrical fields right above the gold surface. The maximum field intensity of nonlift-off grating is about two orders of magnitude larger than that of the lift-off grating. This might account for the larger enhancement for the nonlift-off grating, because the nonlift-off grating was covered with a corrugated continuous gold thin film, which supports the excitation of propagating SPP [32]. The excitation of SPP on the gold surface was responsible for the largely enhanced field. However, the gold thin film might have quenched the generated signals, so the measured signal was not as high as simulated.

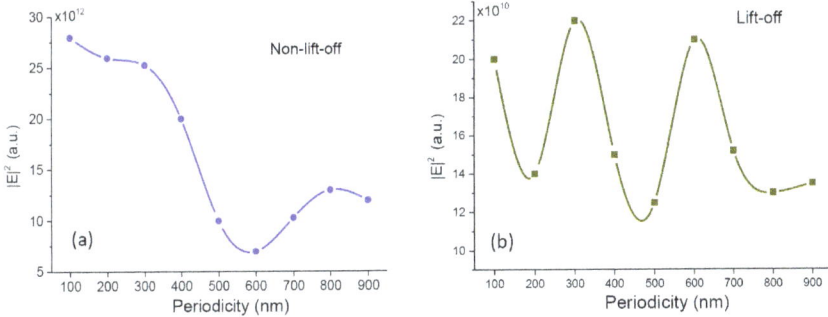

Figure 7. Calculated electrical field intensity on the gold surface of (**a**) non-lift-off and (**b**) lift-off nanogratings. The medium above the gold surface is water. The field strength is averaged over the entire gold surface of a unit cell.

About the periodicity dependence, the simulation agrees well with the experimental results for periods longer than 300 nm. The nonlift-off grating had a minimum at a period of 600 nm as in the experimental result. The lift-off grating had maximums at 300 and 600 nm, as seen in the experimental data. The simulation confirms the LSPR-induced enhancement for the lift-off nanogratings.

For grating periods smaller than 200 nm, the field strength seems to be larger for nonlift-off structures. The electrical field for lift-off grating was also larger at 100 nm. This is because the electrical field is significantly enhanced at small features due to LSPR. However, it is difficult to make nanogratings with a period of 100 nm in real life. A period of 100 nm means the feature size is only tens of nanometers. It is challenging to make a uniform substrate with such small features. The failure rate of lift-off is also higher. As a rule of thumb, better periodicity of the grating results in a better SPP excitation and higher SERS signal intensity [33]. It was the poor homogeneity of our 100 nm-period gratings that resulted in low surface plasmon excitation efficiency. Nevertheless, as seen in the simulation results, one does not gain much field strength even with a 100-nm grating. From the viewpoint of real-world applications, it is not worth the effort to struggle with the fabrication of small gratings.

As mentioned previously, the strong field on the nonlift gratings was due to the excitation of SPP on the corrugated continuous gold thin film. The further enhancement at smaller periods was due to coupling between surface plasmon polariton and localized surface plasmons [30]. This coupling has been shown experimentally and applied for SERS sensing with a uniform gold thin layer beneath gold nanodisks and gratings [30,34]. To excite SPP on the interface, the excitation angle must satisfy the SP dispersion relation as given in Equation (3):

$$k_0 sin\theta \pm K_G = \frac{2\pi}{\lambda_0} sin\theta \pm m\frac{2\pi}{\Lambda} = k_{sp} \qquad (3)$$

where k_0 is the free space wave vector, θ is the incident angle, K_G is the Bragg vector supported by the grating, and m is an integer, which refers to the order of diffraction. k_{sp} is the surface plasmon wave vector, which is given in Equation (1).

Figure 8 plots the dispersion relation of Equation (3) on water–gold and PMMA–gold interfaces respectively. The NA of our objective is 0.5, which corresponds to a maximum excitation and collection angle of 30 degrees. With the large excitation angle, multiple modes of SPPs can be excited with the help of nanogratings as long as the excitation angle and grating orientation match the dispersion relation [35,36]. As seen in Figure 8a, for the water–gold interface, the resonance angle for gratings of period larger than 500 nm was larger than 30 degrees. The period of 500 nm was right on the margin of the NA, so we saw a quick drop in the excitation efficiency at 500 nm period nanograting in the simulation. In the protein-binding experiment, the difference in the signals for Raman and fluorescence at 500 nm period might be explained by the dispersion plot as well. For SERS signals, because of the lower excitation efficiency at the period of 500 nm, the Raman intensity was lower as seen in Figure 6b. However, for the fluorescence intensity, because a majority of fluorescence photons had a wavelength longer than 550 nm (emission peak of the fluorophore), their resonance occurred at an angle smaller than 30 degrees, as seen in Figure 8a. These photons may have been coupled as SPP surface waves and re-emitted at the resonance angle where the phase mating condition was satisfied [37]. The out-coupled fluorescence photons emit at a smaller angle than can be collected by the lens, thus a higher fluorescence signal was obtained.

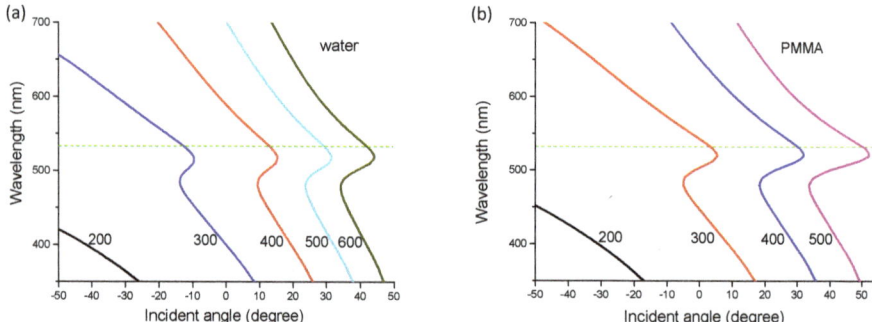

Figure 8. Plot of dispersion relation on (**a**) gold–water and (**b**) gold–PMMA interfaces. The numbers next to the curves are the grating period in nanometer. The green horizontal dashed line indicates the excitation wavelength of 532 nm.

For the PMMA–gold interface, as shown in Figure 8b, the resonance angle for 500 nm is outside the NA of the lens. Therefore we observed an intensity minimum at a period of 500 nm for the nonlift-off PMMA–coated nanograting as shown in Figure 5a. The SERS and fluorescence intensities increased again at longer periods because of the satisfaction of second order (m = 2) phase-matching conditions. For example, the second order dispersion relation of 600 nm coincided with the 300 nm period and the 700 nm coincided with 350 nm. Their resonance angles would be in the range of 0–20 degrees, which can be excited and collected with our lens. However, the efficiency of second order was smaller than the first order.

The above discussion elucidates the periodicity dependence of SERS and fluorescence intensities on the nanogratings. Because the continuous corrugated gold thin film supported the excitation of SPP, the electrical field intensity was two orders of magnitude larger on the non-lift nanograting. Due to the matching of the dispersion relation of SPP, we had the highest SERS and fluorescence signals at periods between 200–400 nm on the non-lift nanogratings. For 100 or 200 nm period grating, the dispersion relation did not directly support the excitation of SPP. However, because the corrugated small periods had Fourier components of longer periods, it could still support the excitation of SPP. The coupling of SPP and LSP at the nanosized grating grooves greatly enhanced the electromagnetic field strength [38]. Because the substrates were fabricated by E-beam lithography, our substrates had excellent homogeneity. Due to the uniform field distribution, the SERS signal was reproducible and robust. Because no lift-off process was needed, the fabrication process was easy and hassle-free. Therefore it is suitable for industrial mass production. It is also possible to make two-dimensional (2D) structures to further improve the enhancement efficiency as demonstrated in several reports [21,30,39]. For solely SERS sensing applications, a longer wavelength excitation laser of 785 nm may be employed, which resonates at a longer period. It is easier to fabricate nanogratings of a larger period. Further improvement of the enhancement factor is possible by decorating the grating surface with nanoparticles and optimizing the geometry and materials [28]. Our simple method allows the generation of SERS-active substrate in a short time with high efficiency. We envision this method to be useful as a cost-effective technique for the production of large-area, SERS-active substrates for real-world applications.

4. Conclusions

In conclusion, we proposed a simplified E-beam lithography method to fabricate SERS-active substrates. By eliminating the lift-off process, it is possible to make large-area, SERS-active substrates with high efficiency. The fabricated nanograting substrates were highly uniform and thus exhibited reproducible and robust SERS signals. We observed the highest SERS intensity on nanogratings of periods 200–400 nm. The enhancement had an obvious periodicity dependence. By investigating the fluorescence enhancement on the lift-off and nonlift-off nanogratings, we found the enhancement on the two structures is due to distinct plasmonic effects. The excitation of SPP was responsible for the large enhancement and SERS effect on the substrate. Our numerical simulations agreed nicely with the experimental results and indicated that further increasing of SERS efficiency is possible by the coupling of SPP and LSPR. We have demonstrated using this substrate for the detection of protein-specific binding. Our method has the potential to fabricate large-area SERS-active substrates based on the relatively matured lithographic technique in a shorter period. The proposed SERS-active substrate can be readily employed as a routine molecular sensing element for a wide range of applications, such as environmental pollutant surveillance and immunoassays.

Author Contributions: Conceptualization, Y.-C.C.; Methodology, Y.-C.C.; Experiment, Y.-C.C. and B.-H.H.; Simulation, Y.-C.C. and T.-H.L.; Validation, Y.-C.C.; Formal analysis, Y.-C.C.; Investigation, Y.-C.C.; Writing, Y.-C.C.; Visualization, Y.-C.C.; Supervision, Y.-C.C.; Project administration, Y.-C.C.; Funding acquisition, Y.-C.C. All authors have read and agreed to the published version of the manuscript.

Funding: This research was funded by Ministry of Science and Technology of Taiwan, grant numbers MOST 106-2221-E-018-021, MOST 107-2221-E-018-006, and MOST 108-2221-E-018-016.

Acknowledgments: The authors would like to express their gratitude to the support from the Ministry of Science and Technology of Taiwan and the technical assistance from the nanofabrication facility of National Chang-Hua University of Education. Y.-C.C. would also like to appreciate the many fruitful discussions with Yu-Ju Hung, Kuo-Ping Chen, Kerwin Wang and Yang-Wei Lin and their technical assistances.

Conflicts of Interest: The authors declare no conflict of interest.

References

1. Fleischmann, M.; Hendra, P.; McQuillan, A. Raman Spectra of Pyridine Adsorbed at a Silver Electrode. *Chem. Phys. Lett.* **1974**, *26*, 163–166. [CrossRef]
2. Ding, S.Y.; Yi, J.; Li, J.F.; Ren, B.; Wu, D.Y.; Panneerselvam, R.; Tian, Z.Q. Nanostructure-based plasmon-enhanced Raman spectroscopy for surface analysis of materials. *Nat. Rev. Mater.* **2016**, *1*, 1–16. [CrossRef]
3. Moskovits, M. Surface roughness and the enhanced intensity of Raman scattering by molecules adsorbed on metals. *J. Chem. Phys.* **1978**, *69*, 4159–4161. [CrossRef]
4. Nie, S.; Emory, S.R. Probing single molecules and single nanoparticles by surface-enhanced Raman scattering. *Science* **1997**, *275*, 1102–1106. [CrossRef] [PubMed]
5. Kneipp, K.; Wang, Y.; Kneipp, H.; Perelman, L.T.; Itzkan, I.; Dasari, R.R.; Feld, M.S. Single molecule detection using surface-enhanced Raman scattering (SERS). *Phys. Rev. Lett.* **1997**, *78*, 1667. [CrossRef]
6. Wang, A.X.; Kong, X. Review of recent progress of plasmonic materials and nano-structures for surface-enhanced Raman scattering. *Materials* **2015**, *8*, 3024–3052. [CrossRef] [PubMed]
7. McNay, G.; Eustace, D.; Smith, W.E.; Faulds, K.; Graham, D. Surface-enhanced Raman scattering (SERS) and surface-enhanced resonance Raman scattering (SERRS): A review of applications. *Appl. Spectrosc.* **2011**, *65*, 825–837. [CrossRef]
8. Sharma, B.; Frontiera, R.R.; Henry, A.-I.; Ringe, E.; van Duyne, R.P. SERS: Materials, applications, and the future. *Mater. Today* **2012**, *15*, 16–25. [CrossRef]
9. Luo, S.-C.; Sivashanmugan, K.; Liao, J.-D.; Yao, C.-K.; Peng, H.-C. Nanofabricated SERS-active substrates for single-molecule to virus detection in vitro: A review. *Biosens. Bioelectron.* **2014**, *61*, 232–240. [CrossRef]
10. Yuan, H.; Liu, Y.; Fales, A.M.; Li, Y.L.; Liu, J.; Vo-Dinh, T. Quantitative surface-enhanced resonant Raman scattering multiplexing of biocompatible gold nanostars for in vitro and ex vivo detection. *Anal. Chem.* **2013**, *85*, 208–212. [CrossRef]
11. Kalachyova, Y.; Mares, D.; Lyutakov, O.; Kostejn, M.; Lapcak, L.; Svorcik, V. Surface plasmon polaritons on silver gratings for optimal SERS response. *J. Phys. Chem. C* **2015**, *119*, 9506–9512. [CrossRef]
12. Mandal, P.; Gupta, P.; Nandi, A.; Ramakrishna, S.A. Surface enhanced fluorescence and imaging with plasmon near-fields in gold corrugated gratings. *J. Nanophotonics* **2012**, *6*, 063527. [CrossRef]
13. Barbillon, G.; Bijeon, J.-L.; Plain, J.; de la Chapelle, M.L.; Adam, P.-M.; Royer, P. Electron beam lithography designed chemical nanosensors based on localized surface plasmon resonance. *Surf. Sci.* **2007**, *601*, 5057–5061. [CrossRef]
14. Yue, W.; Wang, Z.; Yang, Y.; Chen, L.; Syed, A.; Wong, K.; Wang, X. Electron-beam lithography of gold nanostructures for surface-enhanced Raman scattering. *J. Micromech. Microeng.* **2012**, *22*, 125007. [CrossRef]
15. Lin, Y.-Y.; Liao, J.-D.; Ju, Y.-H.; Chang, C.-W.; Shiau, A.-L. Focused ion beam-fabricated Au micro/nanostructures used as a surface enhanced Raman scattering-active substrate for trace detection of molecules and influenza virus. *Nanotechnology* **2011**, *22*, 185308. [CrossRef]
16. Barbillon, G.; Hamouda, F.; Held, S.; Gogol, P.; Bartenlian, B. Gold nanoparticles by soft UV nanoimprint lithography coupled to a lift-off process for plasmonic sensing of antibodies. *Microelectron. Eng.* **2010**, *87*, 1001–1004. [CrossRef]
17. Fan, M.; Andrade, G.F.; Brolo, A.G. A review on the fabrication of substrates for surface enhanced Raman spectroscopy and their applications in analytical chemistry. *Anal. Chim. Acta* **2011**, *693*, 7–25. [CrossRef]
18. Guillot, N.; de la Chapelle, M.L. Lithographied nanostructures as nanosensors. *J. Nanophotonics* **2012**, *6*, 064506. [CrossRef]
19. Halas, N.J.; Lal, S.; Chang, W.-S.; Link, S.; Nordlander, P. Plasmons in strongly coupled metallic nanostructures. *Chem. Rev.* **2011**, *111*, 3913–3961. [CrossRef]

20. Le Ru, E.; Etchegoin, P. *Principles of Surface-Enhanced Raman Spectroscopy: And Related Plasmonic Effects*; Elsevier: Oxford, UK, 2008.
21. Mandal, P.; Ramakrishna, S.A. Dependence of surface enhanced Raman scattering on the plasmonic template periodicity. *Opt. Lett.* **2011**, *36*, 3705–3707. [CrossRef]
22. Hwang, E.; Smolyaninov, I.I.; Davis, C.C. Surface plasmon polariton enhanced fluorescence from quantum dots on nanostructured metal surface. *Nano Lett.* **2010**, *10*, 813–820. [CrossRef] [PubMed]
23. Guo, S.-H.; Heetderks, J.J.; Kan, H.-C.; Phaneuf, R.J. Enhanced fluorescence and near-field intensity for Ag nanowire/nanocolumn arrays: Evidence for the role of surface plasmon standing waves. *Opt. Express* **2008**, *16*, 18417–18425. [CrossRef] [PubMed]
24. Yakubovsky, D.I.; Arsenin, A.V.; Stebunov, Y.V.; Fedyanin, D.Y.; Volkov, V.S. Optical constants and structural properties of thin gold films. *Opt. Express* **2017**, *25*, 25574–25587. [CrossRef] [PubMed]
25. Zhao, J.; Cheng, Y.; Shen, H.; Hui, Y.Y.; Wen, T.; Chang, H.-C.; Gong, Q.; Lu, G. Light Emission from Plasmonic Nanostructures Enhanced with Fluorescent Nanodiamonds. *Sci. Rep.* **2018**, *8*, 3605. [CrossRef] [PubMed]
26. Kubin, R.F.; Fletcher, A.N. Fluorescence quantum yields of some rhodamine dyes. *J. Lumin.* **1982**, *27*, 455–462. [CrossRef]
27. Chang, Y.-C.; Lu, Y.-C.; Hung, Y.-J. Controlling the nanoscale gaps on silver Island film for efficient surface-enhanced Raman spectroscopy. *Nanomaterials* **2019**, *9*, 470. [CrossRef]
28. Kalachyova, Y.; Mares, D.; Jerabek, V.; Zaruba, K.; Ulbrich, P.; Lapcak, L.; Svorcik, V.; Lyutakov, O. The effect of silver grating and nanoparticles grafting for LSP–SPP coupling and SERS response intensification. *J. Phys. Chem. C* **2016**, *120*, 10569–10577. [CrossRef]
29. Hessel, A.; Oliner, A. A new theory of Wood's anomalies on optical gratings. *Appl. Opt.* **1965**, *4*, 1275–1297. [CrossRef]
30. Bryche, J.-F.; Gillibert, R.; Barbillon, G.; Gogol, P.; Moreau, J.; de la Chapelle, M.L.; Bartenlian, B.; Canva, M. Plasmonic enhancement by a continuous gold underlayer: Application to SERS sensing. *Plasmonics* **2016**, *11*, 601–608. [CrossRef]
31. Galarreta, B.C.; Norton, P.R.; Lagugne-Labarthet, F. SERS detection of streptavidin/biotin monolayer assemblies. *Langmuir* **2011**, *27*, 1494–1498. [CrossRef]
32. Zayats, A.V.; Smolyaninov, I.I. Near-field photonics: Surface plasmon polaritons and localized surface plasmons. *J. Opt. A Pure Appl. Opt.* **2003**, *5*, S16. [CrossRef]
33. Yang, J.; Li, J.; Gong, Q.; Teng, J.; Hong, M. High aspect ratio SiNW arrays with Ag nanoparticles decoration for strong SERS detection. *Nanotechnology* **2003**, *25*, 465707. [CrossRef] [PubMed]
34. Chu, Y.; Banaee, M.G.; Crozier, K.B. Double-resonance plasmon substrates for surface-enhanced Raman scattering with enhancement at excitation and stokes frequencies. *ACS Nano* **2010**, *4*, 2804–2810. [CrossRef] [PubMed]
35. Hung, Y.-J.; Smolyaninov, I.I.; Davis, C.C.; Wu, H.-C. Fluorescence enhancement by surface gratings. *Opt. Express* **2006**, *14*, 10825–10830. [CrossRef]
36. Byrne, D.; Duggan, P.; McDonagh, C. Controlled surface plasmon enhanced fluorescence from 1D gold gratings via azimuth rotations. *Methods Appl. Fluoresc.* **2017**, *5*, 015004. [CrossRef]
37. Lakowicz, J.R.; Malicka, J.; Gryczynski, I.; Gryczynski, Z. Directional surface plasmon-coupled emission: A new method for high sensitivity detection. *Biochem. Biophys. Res. Commun.* **2003**, *307*, 435–439. [CrossRef]
38. Genevet, P.; Tetienne, J.P.; Gatzogiannis, E.; Blanchard, R.; Kats, M.A.; Scully, M.O.; Capasso, F. Large enhancement of nonlinear optical phenomena by plasmonic nanocavity gratings. *Nano Lett.* **2010**, *10*, 4880–4883. [CrossRef]
39. Cui, X.; Tawa, K.; Kintaka, K.; Nishii, J. Enhanced fluorescence microscopic imaging by plasmonic nanostructures: From a 1D grating to a 2D nanohole array. *Adv. Funct. Mater.* **2010**, *20*, 945–950. [CrossRef]

© 2020 by the authors. Licensee MDPI, Basel, Switzerland. This article is an open access article distributed under the terms and conditions of the Creative Commons Attribution (CC BY) license (http://creativecommons.org/licenses/by/4.0/).

Article

Surface Enhanced Raman Scattering on Regular Arrays of Gold Nanostructures: Impact of Long-Range Interactions and the Surrounding Medium

Iman Ragheb [1,†], Macilia Braïk [2,†], Stéphanie Lau-Truong [1,†], Abderrahmane Belkhir [2,†], Anna Rumyantseva [3,†], Sergei Kostcheev [3,†], Pierre-Michel Adam [3,†], Alexandre Chevillot-Biraud [1,†], Georges Lévi [1,†], Jean Aubard [1,†], Leïla Boubekeur-Lecaque [1,†] and Nordin Félidj [1,*,†]

1. Université de Paris, Laboratoire ITODYS, CNRS, F-75006 Paris, France; imanraghebomar@gmail.com (I.R.); stephanie.lau@univ-paris-diderot.fr (S.L.-T.); alexandre.chevillot@univ-paris-diderot.fr (A.C.-B.); georges.levi@univ-paris-diderot.fr (G.L.); jean.aubard@univ-paris-diderot.fr (J.A.); leila.boubekeur@univ-paris-diderot.fr (L.B.-L.)
2. Université Mouloud Mammeri de Tizi-Ouzou, LPCQ, BP 17 RP, 15000 Tizi-Ouzou, Algeria; massiliabraik@gmail.com (M.B.); belabd2000@yahoo.fr (A.B.)
3. Charles Delaunay Institute, P2MN Department, University of Technology of Troyes, CS 42060, 10004 Troyes, France; anna.rumyantseva@utt.fr (A.R.); sergei.kostcheev@utt.fr (S.K.); pierre_michel.adam@utt.fr (P.-M.A.)
* Correspondence: nordin.felidj@univ-paris-diderot.fr
† These authors contributed equally to this work.

Received: 7 October 2020; Accepted: 30 October 2020; Published: 4 November 2020

Abstract: Long-range interaction in regular metallic nanostructure arrays can provide the possibility to manipulate their optical properties, governed by the excitation of localized surface plasmon (LSP) resonances. When assembling the nanoparticles in an array, interactions between nanoparticles can result in a strong electromagnetic coupling for specific grating constants. Such a grating effect leads to narrow LSP peaks due to the emergence of new radiative orders in the plane of the substrate, and thus, an important improvement of the intensity of the local electric field. In this work, we report on the optical study of LSP modes supported by square arrays of gold nanodiscs deposited on an indium tin oxide (ITO) coated glass substrate, and its impact on the surface enhanced Raman scattering (SERS) of a molecular adsorbate, the mercapto benzoic acid (4-MBA). We estimated the Raman gain of these molecules, by varying the grating constant and the refractive index of the surrounding medium of the superstrate, from an asymmetric medium (air) to a symmetric one (oil). We show that the Raman gain can be improved with one order of magnitude in a symmetric medium compared to SERS experiments in air, by considering the appropriate grating constant. Our experimental results are supported by FDTD calculations, and confirm the importance of the grating effect in the design of SERS substrates.

Keywords: localized surface plasmon; surface enhanced Raman scattering; grating effect; gold nanodisks; Rayleigh anomaly

1. Introduction

Over the two last decades, metallic nanostructures led to a lot of research in nano-optics, thanks to their unique plasmonic properties [1]. These properties are connected to localized surface plasmon (LSP) resonances associated to collective oscillations of conductive electrons at the surface of the nanoparticles

(NPs) [2]. The LSP wavelength depends on the geometrical parameters of the NPs, the chemical composition of the metallic NPs, the inter-particles distance and the surrounding medium [3,4]. In addition, these optical proprieties are characterized by a strong extinction in the far-field in the visible and near-infrared range (mainly for gold and silver), and a strong electric field enhancement in the near-field of the nanostructures [5].

Depending on the distance between nanoparticles, two coupling modes can be considered: a short-range coupling in the near-field of the particles and a long-range coupling [6–9]. The short-range coupling occurs when the separation distance d is much smaller than the optical wavelength λ (typically d smaller than 10 nm). The particles are thus treated as dipoles interacting through their near-field [10]. Near-field coupling results from the Coulomb interaction between the surface charges on particles and becomes stronger when the areas presenting a high charge density are close to each other, and increases when the distance between nanoparticles is reduced. This type of coupling exhibits large charge dipoles particularly in the gap between nanoparticles, leading to strong local fields compared to the case of isolated nanoparticles [11]. As a result, the separation distance strongly affects the optical response of the system. For instance, when distance between two nanodiscs decreases, the LSP resonance is red-shifted due to the decreasing of the restoring force for single nanoparticles (for a polarization parallel to the main axe of the NP dimer) [12]. As a consequence, the splitting energy between the new hybridized modes is increased and the coupling becomes stronger. In addition, this coupling has a strong impact on the near-field of the nanoparticles. Compared to a single nanoparticle, a dimer of nanoparticles exhibits a higher electric field enhancement due to the dipoles interaction between the plasmon modes, mainly located in the gap between the nanoparticles (called hot-spots) [13,14].

A long-range interaction in regular metallic nanostructure arrays can also provide the possibility to manipulate their optical properties [15,16]. When assembling the nanoparticles in an array, interactions between nanoparticles can result in long-range interactions, for specific inter-particle distances (grating constant) [17,18]. As a result, the optical response exhibits a narrow LSP peak due to the emergence of radiative orders in the plane of the substrate [19–22]. Plasmonic nanostructures arranged in regular arrays support lattice (or collective) plasmon modes, and the interference between localized surface plasmon (LSP) and the so-called Rayleigh anomaly leads to a reduced linewidth of the resonance, and thus, an important improvement of its quality factor [23,24]. Such effect finds applications in non-linear optics [25], molecular sensing [26], plasmon-based lasers [27], surface enhanced fluorescence [28] and surface enhanced Raman scattering (SERS) [29–31].

Only a very few works related to long-range interactions have been dedicated to this field. The SERS effect originates mainly from an electromagnetic enhancement mechanism consecutive to the excitation of localized surface plasmon (LSP) of metallic nanoparticles, and takes place for molecules (including at very low concentration) close to the surface of metallic particles, provided that the laser line wavelength is close to the maximum of LSP resonance [32,33]. In particular, molecules adsorbed in the first surface layer display the largest Raman enhancement factors (REF). Taking into account both enhanced fields, the average Raman gain $< G >$ can be expressed as [34]:

$$< G > = \left\langle |A(\nu_{exc})|^2 \times |A(\nu_R)|^2 \right\rangle \qquad (1)$$

where $A(\nu_{exc})$ is the local electric-field enhancement factor at the incident frequency ν_{exc}, and $A(\nu_R)$ is the corresponding factor at the Raman frequency ν_R. Most of the time, $\langle G \rangle$ is averaged over the surface area of the particles, in order to estimate the Raman gain. In general, $\langle G \rangle$ is approximated by assuming that $A(\nu_{exc})$ and $A(\nu_R)$ are identical; hence, $\langle G \rangle$ can be rewritten $\langle G \rangle \sim |A(\nu_{exc})|^4$ [35]. This approximation takes advantage of the fact that the LSP width is often large compared to the Stokes shift, except for calibrated samples like lithographic structures, where the LSP band can be narrow [36].

Recently, directional plasmon excitation and SERS studies have been investigated for arrays of gold lines deposited on a gold film [37]. The excitation of the surface plasmon polariton (SPP) takes place either at the metal-air interface or the metal-glass interface leading to the appearance of diffractive modes. Such configuration, although interesting, prevented to estimate easily Raman gains, due to the roughness of the gold film, contributing also the Raman enhancement of the molecular probes. In this work, we report on the optical study of LSP modes supported by square arrays of gold nanodiscs deposited on a indium tin oxyde (ITO) coated glass substrate, and its impact on the SERS response of a molecular adsorbate, the mercapto benzoic acid (4-MBA). We estimated the Raman gain of these molecules, by varying the grating constant and the refractive index of the surrounding medium of the superstrate, from an asymmetric medium (air) to a symmetric medium (oil) with respect to the substrate. We show that the Raman gain can be improved with one order of magnitude in a symmetric medium compared to SERS experiments in air, by considering the appropriate grating constant in accordance with FDTD calculations. They confirm the importance and the impact of the grating effect in the design of SERS substrates.

2. Results and Discussion

Gold nanodiscs arrays (size of 60 × 60 µm^2) were fabricated by electron beam lithography (EBL). The gold nanodiscs height and diameter have been fixed to h = 50 nm and D = 100 nm, respectively. The grating constant (inter-particle distance center-to-center) is varying from Λ = 250 nm to 550 nm. As seen in the Figure 1, the arrays are homogeneous in term of grating constant. Several configurations can be considered depending on the index of the over layer (upper medium).

- The upper medium is air with n = 1 index. This configuration leads to an asymmetric environment since the index of the ITO substrate varies from n = 1.9 to n = 1.7 in the wavelength range.
- The upper medium is water with index n ≃ 1.33 enabling a partial matching with the ITO substrate index.
- The upper medium oil matching index of n ≃ 1.55 leading to a better matching with ITO index.

It has been shown theoretically and experimentally that matched indices improves greatly the grating effect [8,18].

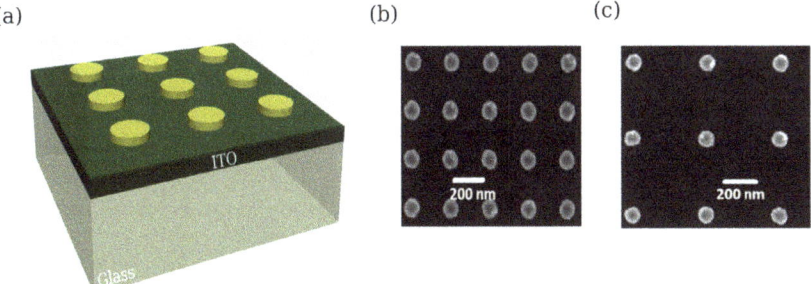

Figure 1. (a) Scheme of the gold nanoparticle (NP) array deposited on an indium tin oxyde (ITO) (thickness 80 nm) coated glass substrate and surrounded with a dielectric of refractive index n. (b,c) SEM images of gold nanodisc square arrays with a diameter of 100 nm and a grating constant Λ of 250 nm and 450 nm, respectively. Height of the discs h = 50 nm.

We first focus our attention on the extinction spectra and SERS experiments on square arrays of gold discs in air, with grating constants varying from Λ = 250 to 550 nm (see Figure 2). For short grating constants, the LSP position is very close to the LSP resonance of isolated nanoparticles. Indeed, no diffracted order is observed in such situation. Figure 2b displays the calculated extinction spectra for a grating constant using the FDTD method. The experimental spectra are in very good qualitative agreement with the calculated ones, although with a smaller full width at half maximum (FWHM) and slightly blue-shifted for the calculated ones, due to the fact that, in the calculations, the nanoscale surface roughness (NSR) of the gold discs was not taken into account [38,39]. When the grating constant is increased, a significant red-shift of the LSP is expected, as well as a reduced FWHM. This optical behavior is confirmed by the calculated extinction spectra when varying the grating constant, as displayed in the Figure 2b. The extinction spectra are attributed to collective LSP resonances (so-called lattice modes).

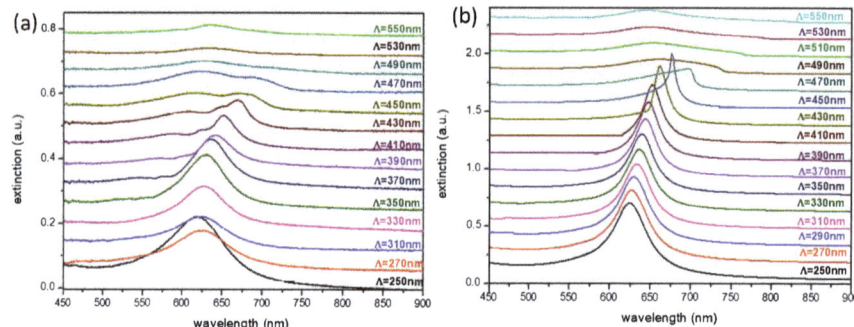

Figure 2. (a) Experimental and (b) calculated extinction spectra in air (in transmission, light at normal incidence and polarized along the x-axis) for square arrays of gold nanodiscs with a diameter D = 100 nm. The grating constant is varying from Λ = 250 to 550 nm with steps of 20 nm. The height of nanodiscs is fixed to h = 50 nm. The calculations have been obtained by the FDTD method.

According to the grating theory, for a grating constant of Λ_c with a refractive index of the substrate n_{sub}, the position of the Rayleigh anomaly is given by $\Lambda_c = \lambda/n_{sub}$, for an illumination at normal incidence. In the range of LSP wavelengths and grating constants considered in Figure 2, the Rayleigh anomaly can be excited, corresponding to (±1,0) diffraction order in the plane of the substrate. Therefore, when the lattice wavelength is close to the positions of the Rayleigh anomaly, a reduced FWHM is observed. This behavior is due to a strong coupling between the LSP mode and the Rayleigh anomaly, which is observed for a critical grating constant Λ_c = 410 nm in our experiments (Figure 2a). For such grating constant (Λ = 410 nm), a reduced FWHM with a quality factor of Q_c = 18.75 is measured, and higher compared to the case of Λ = 270 nm, for which Q = 8.68 (the quality factor is defined as $Q = \omega/\Delta\omega$, where ω and $\Delta\omega$ are the resonance frequency and the resonance width at half-max, respectively).

Since the quality factor Q is increasing, the near-field intensity is expected to also increase. In the Figure 3, we present the FDTD calculation of the intensity of the local electric field (calculated at the maximum of the lattice mode wavelength) versus the grating constant. It can be seen that the maximum of intensity is obtained for Λ = 430 nm, corresponding to the calculated critical grating constant Λ_c, where a strong long-range coupling occurs. A second maximum at $\Lambda \sim 620$ nm, with a much lower intensity, is observed, and attributed to a grating order (0,±1) in air. The slight difference between the experimental and calculated Λ_c comes from the fact that in the calculations, the NSR is not taken into account [38,39].

We thus expect that the choice of the grating constant, in the context of SERS measurements, will impact significantly the Raman enhancement factor. However, in order to be able to compare the experimental SERS measurements versus the grating constant with the calculated REF, one has to take into account the calculated REF defined as $REF_{calc} = |E(\omega_{exc})|^2 * |E(\omega_{RS})|^2$ (ω_{exc} and ω_{RS} correspond to the laser excitation and Raman (Stokes side) angular frequency, respectively. We thus plotted the REF_{calc} versus the grating constant, by considering the two Raman lines at 1074 and 1585 cm^{-1}, corresponding to characteristic Raman bands of 4-MBA molecules, for an incident wavelength at 633 nm (excitation line used in our experiments).

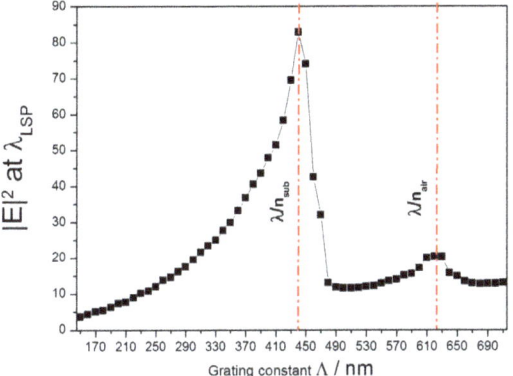

Figure 3. Calculated intensity of the local electric field at the maximum of the LSP mode wavelength versus the grating constant. Calculations are made by the FDTD method for gold discs (D = 100 nm and h = 50 nm).

As displayed in Figure 4, the maximum of REF_{calc} is expected to be maximum at a smaller grating constant ($\Lambda = 370$ nm), compared to the maximum of intensity measured at λ_{LSP} (at $\Lambda = 430$ nm). The REF_{calc} is also compared to $|E(\omega_{exc})|^4$ at $\lambda_{exc} = 633$ nm, which corresponds to an approximation often used for the estimation of the REF. It is seen that the maximum of $|E(\omega_{exc})|^4$ (for an incident wavelength at 633 nm) is obtained for a smaller grating constant, compared to the maximum of REF_{calc}. This difference was expected since one has to take into account the enhancement factor $|E(\omega_{RS})|^2$ at λ_{RS}, red-shifted compared to the enhancement factor $|E(\omega_{exc})|^2$ at λ_{exc}.

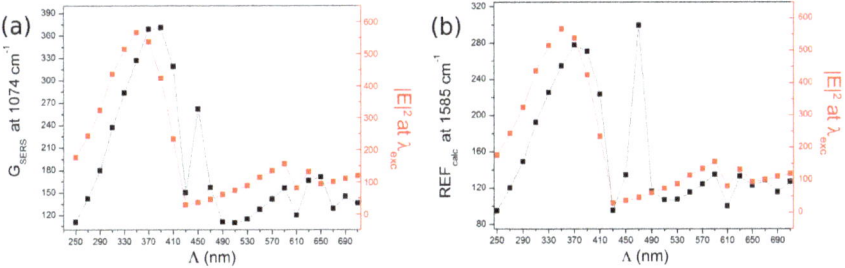

Figure 4. Calculated Raman enhancement factor (REF_{calc}) versus the grating constant for the Raman line at 1074 cm^{-1} (**a**) and 1585 cm^{-1} (**b**). The REFs are superposed with $|E(\omega_{exc})|^4$, calculated at 633 nm. The calculation are made for gold nanodiscs arrays with a diameter of D = 100 nm. The height of nanodiscs is h = 50 nm.

Finally, a "sharp" maximum is also observed at Λ = 450 nm for the 1074 cm^{-1} Raman line, and at Λ = 470 nm for the 1585 cm^{-1} Raman line. This difference in grating constant is due to the fact that the Raman emissions are located at different wavelengths. For instance, the Raman emission at 1585 cm^{-1} is more predominant for a higher grating constant since it corresponds to an LSP wavelength more red-shifted compared to those of a smaller Λ. These two additional maxima are attributed to the fact that the maximum of the LSP is precisely located at half way between the excitation line and the Raman lines ($\lambda_{LSP} = (\lambda_{exc} + \lambda_{RS})/2$), leading to an optimized Raman gain, as demonstrated in the reference [36].

In the following, we investigate experimentally the impact of the grating constant on the Raman gain, by considering as the superstrate, the air. The spontaneous Raman spectrum was first characterized for a 0.5 M 4-MBA in a DMSO solution (Figure 5a). The Raman signature is mainly characterized by two intense Raman bands of 4-MBA located at 1074 and 1585 cm^{-1} (spectrum a), associated to CH out-phase bonding and to C=C symmetric stretching vibrations, respectively. The adsorption of the molecules onto monolayers is crucial in order to estimate experimentally the number of molecules contributing to SERS, and thus, the Raman enhancement factor (spectrum b, Figure 5a). In order to verify that 4-MBA molecules form monolayers onto the surface of gold nanoparticles, we recorded the SERS spectra of 4-MBA molecules on a gold nanodisc array (D = 95 nm, h = 50 nm, Λ = 320 nm), with different incubation times (molecular concentration of 10^{-4} M). Figure 5b shows the intensity of the Raman bands at 1074 cm^{-1} and 1585 cm^{-1} versus the incubation time adsorbed on a gold nanodiscs array. The SERS intensity increases and reaches its maximum after $\sim$50 s of incubation time. This intensity remains constant when the incubation time is increased. This result allows us to conclude that a monolayer of 4-MBA molecules is formed for an incubation time of 40–50 s. In the following SERS experiments, we will use an incubation time of 5 min in a solution of 10^{-4} M, in order to insure that the gold particles are fully covered by a monolayer of 4-MBA molecules.

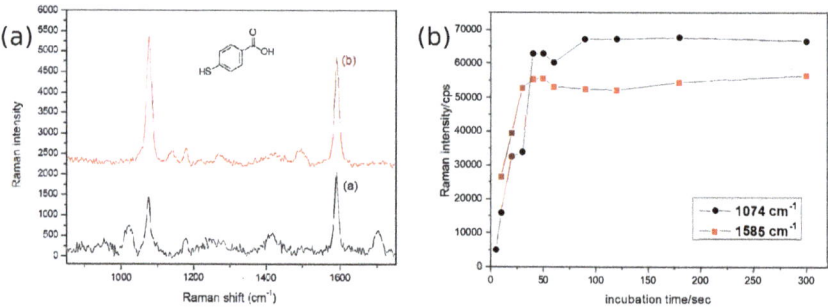

Figure 5. (**a**) Spontaneous Raman spectrum of 4-mercaptobenzoic acid (4-MBA) of (0.5 M) in DMSO solution (spectrum a), and surface enhanced Raman scattering (SERS) spectra of 4-MBA molecules adsorbed on a gold nanodisc array (concentration of 10^{-4} M, sepctrum b). Acquisition conditions for SERS: excitation wavelength λ_{exc} = 632.8 nm, laser power P = 65 µW, acquisition time t = 30 s; (**b**) SERS intensity as a function of incubation time for the Raman bands at 1074 cm^{-1} and 1585 cm^{-1}. Acquisition conditions: excitation wavelength λ_{exc} = 632.8 nm, laser power P = 65 µW, acquisition time t = 30 s.

Since the 4-MBA molecules form a monolayer at the particles surface, it is possible to estimate the order of magnitude of the number of adsorbed molecules (knowing the surface occupied by one molecule), and thus, the Raman enhancement factor (REF). The Raman enhancement factor per molecule is defined as [32,35]:

$$REF = \frac{I_{SERS}/N_{SERS}}{I_{Ref}/N_{Ref}} \quad (2)$$

with $N_{SERS} = (N \times S_{NP})/S_{mol}$ and $N_{Ref} = C \times V_{eff} \times N_A$.

In Equation (2), I_{SERS} corresponds to the integrated intensity of the Raman bands of the 4-MBA molecules, I_{Ref} the integrated Raman intensity corresponding to the spontaneous Raman spectrum recorded for a 0.5 M 4-MBA in a DMSO solution. N_{SERS} is the number of molecules occupied in the laser spot surface, N is the number of metallic nanoparticles under the laser spot area, S_{NP} is the surface occupied by one nanoparticle and S_{mol} is the surface occupied by one molecule of 4-MBA equal to 38.3 Å^2. The laser spot surface has thus been estimated to 5 µm^2 for a 100× microscope objective (N.A. 0.65). Therefore, we could estimate N_{SERS} for the different arrays investigated. N_{Ref} is the number of molecules excited in a volume V_{eff} of the laser waist for 0.5 M 4-MBA solution. N_A is the Avogadro number equal to 6.02×10^{23} mol^{-1}. The volume of laser waist is estimated by considering a cone of apex angle defined by the numerical aperture of the microscope objective and the height of the focusing scope. Using a 100× microscope objective with a numerical aperture NA of 0.65, the volume of laser waist is assumed to be 5000 µm^3, leading to a N_{Ref} value of ~1.5×10^{12} molecules. This definition of the Raman enhancement instead of that given in Refs. [32,35] indeed overestimates slightly the gain by not accounting for the molecules on lateral part of the particles. However, we believe that it is more suitable because we are investigating self assembled molecules chemically adsorbed to the gold surface by the sulfur atom; few molecules are adsorbed on the side of the particles and play only a weak role. Indeed, computations show that the electromagnetic field should be weak on this part of the nanodiscs. Therefore, this definition does not change the conclusion concerning the evolution of the average enhancement with the grating constant which was the main goal of this paper.

The SERS signals were recorded at λ_{exc} = 632.8 nm. Using a microscope objective with a 100× magnification, and a numerical aperture of N.A. = 0.9, the estimated zone of excitation was ~5 µm^2. The Figure 6 displays the Raman enhancement factor versus the grating constant for the Raman bands at 1074 and 1585 cm^{-1}. The REF has been measured, on the order of 10^6, in quite good quantitative agreements with recent works [40]. For both Raman bands, we observe that the maximum of REF is obtained for a grating constant of Λ = 330 nm, and not for Λ = 430 nm, for which it was observed a maximum of intensity at λ_{LSP} (Figure 6a,b). The maximum of the experimental REF$_{exp}$ at Λ = 330 nm is slightly different from the calculated REF$_{calc}$ located at Λ = 370 nm (see the Figure 4). This discrepancy is attributed to the fact that the calculated values are extracted from the spectral profile of the near-field intensities, which reflect the calculated extinction spectra. The calculated spectra are slightly shifted compared to the experimental ones, and thus explain why the calculated REF$_{calc}$ is maximum for a slightly higher grating constant. However, the experimental REFs$_{exp}$ are also observed for smaller grating constants than Λ_C, as confirmed by the FDTD calculations (compare Figure 6a,b and Figure 4).

It is noteworthy that the maximum of REF$_{exp}$, obtained for a grating constant at Λ = 330 nm, is not considerably improved, compared to the REF$_{exp}$ measured for other grating constants (Figure 6a,b). Indeed, a factor of two is observed compared to the lowest values of the experimental REF (for instance at Λ = 290 or 490 nm). However, a factor of the same order of magnitude (~3) is also deduced from the calculations, between the maximum REF$_{calc}$ (at Λ = 370 nm) and the minimum REF$_{calc}$ (at Λ = 490 nm). Finally, if a clear maximum of REF$_{exp}$ is observed for Λ = 330 nm, significant fluctuations of the REFs versus the grating constant are observed. The calculated REF$_{calc}$ also displays some fluctuations, although less obvious (Figure 4). This is attributed to the fact that the enhancement factors are not optimized in air.

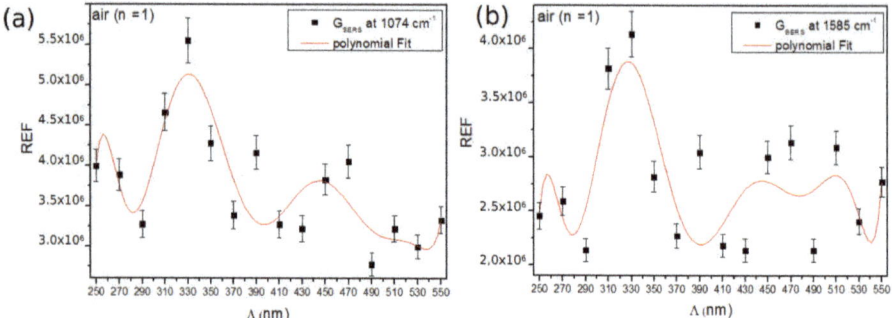

Figure 6. Experimental Raman enhancement factors REF$_{exp}$ as a function of the grating constant for the Raman bands at 1074 cm^{-1} (**a**), and 1585 cm^{-1} (**b**). Acquisition conditions: excitation wavelength λ_{exc} = 632.8 nm, laser power P = 65 µW, acquisition time t = 30 s.

In order to improve the REF, one has to consider a symmetric environment. In other words, the refractive index of the substrate needs to be as close as possible to the refractive index of the superstrate. Indeed, in an asymmetric medium, the radiative patterns by the nanodiscs are mostly scattered inside the substrate. Therefore, the overlap between the grazing diffracted orders and the particle plasmon is limited. In a symmetric environment, the radiative pattern is expected to be symmetric, and thus with a higher coupling, resulting in a strongest near-field intensity [23]. In the following, we thus investigate the impact of the dielectric environment on the far-field and near-field optical response, as well as on the Raman enhancement factors in the context of SERS measurements.

As the superstrate, we considered oil since its refractive index is very close from the substrate (n = 1.51). The Figure 7 displays the extinction spectra of the nanodiscs arrays, recorded in oil, with grating constants varying from Λ = 250 to 490 nm. As expected, a red-shift of the LSP resonances is observed, compared to the ones in air, due to a higher refractive index. The REF$_{calc}$ has been calculated by taking into account the product of the square modulus of the electric field at λ_{exc} = 632.8 nm and the square modulus of the electric field at 1074 cm^{-1} and 1585 cm^{-1} corresponding to $\lambda_{RS,1074}$ = 679 nm and $\lambda_{RS,1585}$ = 704 nm, respectively. Figure 8 displays the calculated REFs$_{calc}$ for the Raman line at 1074 cm^{-1} (Figure 8a) and at 1585 cm^{-1} (Figure 8b). The REFs are also compared to the REFs calculated in air and water (as an intermediate dielectric medium, with a refractive index of n = 1.33).

For both Raman lines, the calculated REFs at maximum are at least 50 times higher in oil compared to air, and 4 times higher, compared to water. Moreover, the maximum of REFs$_{calc}$ in oil corresponds to smaller grating constants (around Λ = 310 nm), compared to the calculated ones in air (around Λ = 370 nm). This can be explained by the fact that the LSP resonances in oil are more red-shifted compared to the excitation line, especially for higher grating constants. Thus, the maximum of REFs$_{calc}$ is expected to be obtained for arrays with smaller grating constants, where the lattice mode wavelength is close to the excitation line at 633 nm. One can note that the maximum of REFs$_{calc}$ is located at Λ = 310 nm for the 1074 cm^{-1} Raman line, and at Λ = 330 nm for the 1585 cm^{-1}. This is attributed to the fact that the 1585 cm^{-1} Raman line corresponds to a higher wavelength compared to that of 1074 cm^{-1} Raman line. Finally, a second maximum of REF, with a lower value at Λ = 530 nm, is observed. This can be attributed to the diffracted ($\pm 1, \pm 1$) order in the substrate plane.

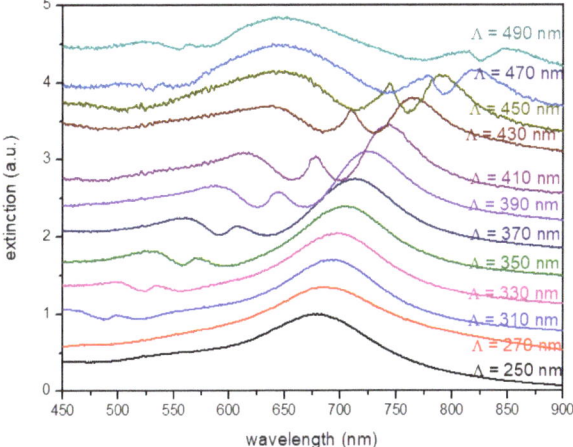

Figure 7. Extinction spectra recorded in oil (in transmission and normal incidence) for square arrays of gold nanodiscs with a diameter D = 100 nm. The grating constant varies from Λ = 250 to 490 nm. The height of nanodiscs is fixed to h = 50 nm.

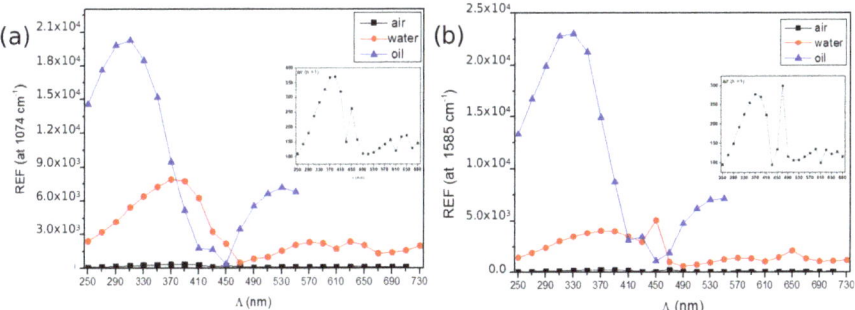

Figure 8. Calculated REFs$_{calc}$ in air (n_a = 1), water (n_w = 1.33) and oil (n_o = 1.518) at: 1074 cm^{-1} (**a**) and 1585 cm^{-1} (**b**) for arrays of gold nanodiscs. Diameter of the disc is D = 100 nm, the height h = 50 nm and grating constant varying from Λ = 250 to 730 nm. The REFs are calculated using the FDTD method.

The experimental SERS measurements in oil have been investigated using a microscope objective (50x, N.A. 0.9), with an excitation line λ_{exc} = 632.8 nm and a laser power of 65 µW. The experimental REFs vary from 2×10^6 (for Λ = 510 nm) to 10^7 (for Λ = 270 nm). Figure 9 compares the experimental REFs versus the grating constant with the calculated REFs in oil. For both Raman lines, there is a good qualitative agreement between the experimental and calculated profile of the REFs versus the grating constant. In particular, the maximum of REF is obtained for a smaller grating constant for the 1074 cm^{-1} Raman line (at Λ = 300 nm), compared to the REF for the 1585 cm^{-1} Raman line (at Λ = 330 nm).

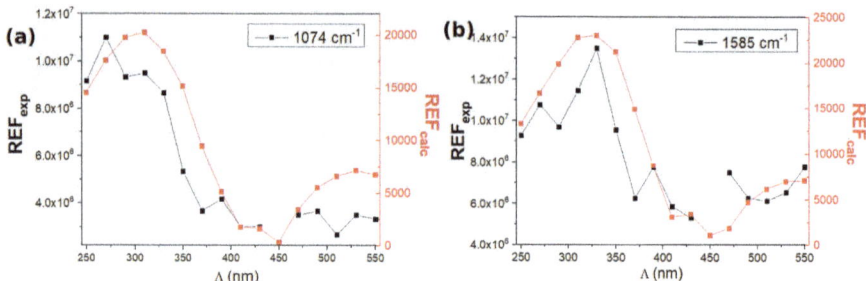

Figure 9. Experimental and calculated REF in oil for the Raman bands at 1074 cm^{-1} (**a**) and 1585 cm^{-1} (**b**).

Note that the REFs$_{exp}$ and REFs$_{calc}$ values are not quantitatively comparable. Although a chemical contribution on the Raman gain may contribute, it could not explain a difference of two orders of magnitude between the experimental and calculated REFs. Such discrepancy has been also pointed out by M. Banaee et al. [40], using the same molecular probe. They noted similar differences between simulation and experimental REF values, attributed to fabrication imperfections. This discrepancy could be explained by the fact that the simulations do not take into account any roughness of the discs surface. Indeed, this roughness is due to the thermal evaporation process before the lift-off step of the EBL fabrication of the samples. Therefore, it has been shown that a significant difference in REF can be observed between smooth (annealed) and roughened (non-annealed) samples. It has been shown that the calculated REF with roughened samples can be ~100 times higher than that for a smooth sample. However, in our experiments, by considering the adequate grating constant in a symmetric environment, we show that the measured REF can be increased by one order of magnitude in comparison to those measured in air, and reaches values of the order of 10^7, which represents high values to detect any molecular probes at very low concentrations.

3. Materials and Methods

Electron beam lithography: The substrates were using an electron beam lithographic system using a ZEISS scanning electronic microscope (SEM) [41]. A 100 nm thick layer of poly-methyl methacrylate electron resist was spin coated on glass substrates with an 80 nm layer of indium tin oxide (ITO). The desired structures were exposed to an electron beam. Chemical development, thermal vacuum coating with gold and a lift-off procedure followed, which led to regular arrays of gold nanodiscs of the desired geometrical parameters, 50 ± 5 nm height and 100 µm diameter on top of the ITO covered glass-substrate. This method allows us to control precisely the nanoparticle size, shape and inter-particle distance between nanoparticles. We have thus the ability to tune the plasmon resonance at any desired wavelength [41,42]. We choose isotropic gold discs in order to avoid any contribution of depolarization effect in Raman measurements, for instance affected by the anisotropy of gold nanorods.

UV–visible absorption spectroscopy: The plasmon bands were characterized by extinction micro-spectroscopy in the range of 400–900 nm. The spectrometer (LOT ORIEL model MS 260i) was coupled to an optical microscope (OLYMPUS BX 51) equipped with a 50× objective (numerical aperture N.A. 0.55).

Raman spectroscopy: The Raman experiments were made using a Jobin–Yvon LABRAM HR 800 Raman spectrometer. The source is an He-Ne laser (632.8 nm), focused on the sample, through a microscope equipped with a 100× objective (Olympus, N.A. 0.8). 4-mercaptobenzoic acid (4-MBA) molecules were used for all the SERS experiments. This molecule is characterized by self-assembled monolayers, when adsorbed on the surface of metallic nanoparticles. Indeed, the adsorption of these molecules takes place via the thiol group on the gold surface.

Finite Difference Time Domain (FDTD) calculations: Finite Difference Time Domain (FDTD) simulations were achieved using a developed 3D-code for the optical proprieties investigation. The code takes into account the periodicity of the structure in x and y directions via Bloch's boundary conditions [43] and the upper and lower semi-infinite media in z direction through perfectly matched layer (PML) conditions of Berenger [44]. The implemented Critical Points Drude model [45] deals with the dispersive nature of gold and ITO using different fitted parameters to match experimental values. The structure is illuminated, with a plane wave, at normal incidence from the substrate. In the near-field, the normalized electric field intensity is calculated in the vicinity of the metallic nanoparticles, while the detector is placed far away from them for far-field simulations of extinction spectra.

4. Conclusions

It has been shown that a long-range coupling within a gold nanodisc array affects significantly the SERS intensity of a molecular probe (4-mercaptobenzoic acid). This type of coupling corresponds to the emergence of a new radiative order in the substrate plane. This interaction is maximum for a critical grating constant Λ_c, when the plasmon mode wavelength is close to the Rayleigh anomaly position. As a consequence, the REF is strongly dependent on the grating constant. The maximum of REF is not obtained for a grating constant corresponding to a maximum of local electric field intensity, but systematically obtained for smaller grating constants regardless the environment. Experimental and theoretical values of the REF display that one needs to consider a symmetric environment, in order to optimize the REF. More importantly, it is demonstrated that the Raman gain of molecular probes can be improved with one order of magnitude in a symmetric medium (in oil) compared to SERS experiments in air, by considering the appropriate grating constant.

Author Contributions: Investigation, I.R., M.B. and A.B.; formal analysis, S.L.-T.; resources, A.R.; data curation, S.K.; methodology, P.-M.A.; software, A.C.-B.; writing—original draft preparation, G.L. and J.A.; writing—review and editing, L.B.-L. and supervision, N.F. All authors have read and agreed to the published version of the manuscript.

Funding: This research was funded by the LabEx SEAM (Science and Engineering for Advanced Materials and devices) of Sorbonne Paris Cité in the frame of the project HOTSPOT.

Conflicts of Interest: The authors declare no conflict of interest.

References

1. Krenn, J.; Weeber, J.; Dereux, A. Nanoplasmonics with Surface Plasmons. In *Advances in Nano-Optics and Nano-Photonics*; Shalaev, V., Kawata, S., Eds.; Elsevier: London, UK, 2007.
2. Maier, S.A. *Plasmonics: Fundamentals and Applications*; Springer Science & Business Media: Berlin/Heidelberg, Germany, 2007.
3. Félidj, N.; Grand, J.; Laurent, G.; Aubard, J.; Levi, G.; Hohenau, A.; Galler, N.; Aussenegg, F.; Krenn, J. Multipolar surface plasmon peaks on gold nanotriangles. *J. Chem. Phys.* **2008**, *128*, 094702. [PubMed]
4. Becker, J.; Trügler, A.; Jakab, A.; Hohenester, U.; Sönnichsen, C. The optimal aspect ratio of gold nanorods for plasmonic bio-sensing. *Plasmonics* **2010**, *5*, 161–167.
5. Hao, E.; Schatz, G.C. Electromagnetic fields around silver nanoparticles and dimers. *J. Chem. Phys.* **2004**, *120*, 357–366. [PubMed]
6. Wang, X.; Gogol, P.; Cambril, E.; Palpant, B. Near-and far-field effects on the plasmon coupling in gold nanoparticle arrays. *J. Phys. Chem. C* **2012**, *116*, 24741–24747.
7. Banaee, M.G.; Crozier, K.B. Gold nanorings as substrates for surface-enhanced Raman scattering. *Opt. Lett.* **2010**, *35*, 760–762.

8. Humphrey, A.; Barnes, W. Plasmonic surface lattice resonances in arrays of metallic nanoparticle dimers. *J. Opt.* **2016**, *18*, 035005.
9. Salerno, M.; Krenn, J.R.; Hohenau, A.; Ditlbacher, H.; Schider, G.; Leitner, A.; Aussenegg, F.R. The optical near-field of gold nanoparticle chains. *Opt. Commun.* **2005**, *248*, 543–549.
10. Rechberger, W.; Hohenau, A.; Leitner, A.; Krenn, J.; Lamprecht, B.; Aussenegg, F. Optical properties of two interacting gold nanoparticles. *Opt. Commun.* **2003**, *220*, 137–141.
11. Aizpurua, J.; Bryant, G.W.; Richter, L.J.; De Abajo, F.G.; Kelley, B.K.; Mallouk, T. Optical properties of coupled metallic nanorods for field-enhanced spectroscopy. *Phys. Rev. B* **2005**, *71*, 235420.
12. Jain, P.K.; El-Sayed, M.A. Plasmonic coupling in noble metal nanostructures. *Chem. Phys. Lett.* **2010**, *487*, 153–164.
13. Haggui, M.; Dridi, M.; Plain, J.; Marguet, S.; Perez, H.; Schatz, G.C.; Wiederrecht, G.P.; Gray, S.K.; Bachelot, R. Spatial confinement of electromagnetic hot and cold spots in gold nanocubes. *ACS Nano* **2012**, *6*, 1299–1307. [CrossRef] [PubMed]
14. Yue, W.; Wang, Z.; Whittaker, J.; Lopez-royo, F.; Yang, Y.; Zayats, A.V. Amplification of surface-enhanced Raman scattering due to substrate-mediated localized surface plasmons in gold nanodimers. *J. Mater. Chem. C* **2017**, *5*, 4075–4084. [CrossRef]
15. Kravets, V.G.; Kabashin, A.V.; Barnes, W.L.; Grigorenko, A.N. Plasmonic surface lattice resonances: A review of properties and applications. *Chem. Rev.* **2018**, *118*, 5912–5951. [CrossRef]
16. Lamprecht, B.; Schider, G.; Lechner, R.; Ditlbacher, H.; Krenn, J.R.; Leitner, A.; Aussenegg, F.R. Metal nanoparticle gratings: influence of dipolar particle interaction on the plasmon resonance. *Phys. Rev. Lett.* **2000**, *84*, 4721. [CrossRef]
17. Yang, X.; Xiao, G.; Lu, Y.; Li, G. Narrow plasmonic surface lattice resonances with preference to asymmetric dielectric environment. *Opt. Express* **2019**, *27*, 25384–25394. [CrossRef]
18. Auguié, B.; Bendana, X.M.; Barnes, W.L.; de Abajo, F.J.G. Diffractive arrays of gold nanoparticles near an interface: Critical role of the substrate. *Phys. Rev. B* **2010**, *82*, 155447. [CrossRef]
19. Kravets, V.; Schedin, F.; Grigorenko, A. Extremely narrow plasmon resonances based on diffraction coupling of localized plasmons in arrays of metallic nanoparticles. *Phys. Rev. Lett.* **2008**, *101*, 087403. [CrossRef]
20. Juodėnas, M.; Tamulevičius, T.; Henzie, J.; Erts, D.; Tamulevičius, S. Surface lattice resonances in self-assembled arrays of monodisperse Ag cuboctahedra. *ACS nano* **2019**, *13*, 9038–9047. [CrossRef]
21. Haynes, C.L.; McFarland, A.D.; Zhao, L.; Van Duyne, R.P.; Schatz, G.C.; Gunnarsson, L.; Prikulis, J.; Kasemo, B.; Käll, M. Nanoparticle optics: the importance of radiative dipole coupling in two-dimensional nanoparticle arrays. *J. Phys. Chem. B* **2003**, *107*, 7337–7342. [CrossRef]
22. Lovera, A.; Gallinet, B.; Nordlander, P.; Martin, O.J. Mechanisms of Fano resonances in coupled plasmonic systems. *ACS Nano* **2013**, *7*, 4527–4536. [CrossRef]
23. Khlopin, D.; Laux, F.; Wardley, W.P.; Martin, J.; Wurtz, G.A.; Plain, J.; Bonod, N.; Zayats, A.V.; Dickson, W.; Gérard, D. Lattice modes and plasmonic linewidth engineering in gold and aluminum nanoparticle arrays. *JOSA B* **2017**, *34*, 691–700. [CrossRef]
24. Ragheb, I.; Braik, M.; Mezeghrane, A.; Boubekeur-Lecaque, L.; Belkhir, A.; Felidj, N. Lattice plasmon modes in an asymmetric environment: from far-field to near-field optical properties. *JOSA B* **2019**, *36*, E36–E41. [CrossRef]
25. Luk'yanchuk, B.; Zheludev, N.I.; Maier, S.A.; Halas, N.J.; Nordlander, P.; Giessen, H.; Chong, C.T. The Fano resonance in plasmonic nanostructures and metamaterials. *Nat. Mater.* **2010**, *9*, 707–715. [CrossRef] [PubMed]
26. Gutha, R.R.; Sadeghi, S.M.; Sharp, C.; Wing, W.J. Biological sensing using hybridization phase of plasmonic resonances with photonic lattice modes in arrays of gold nanoantennas. *Nanotechnology* **2017**, *28*, 355504. [CrossRef] [PubMed]
27. Yang, A.; Hoang, T.B.; Dridi, M.; Deeb, C.; Mikkelsen, M.H.; Schatz, G.C.; Odom, T.W. Real-time tunable lasing from plasmonic nanocavity arrays. *Nat. Commun.* **2015**, *6*, 1–7. [CrossRef]
28. Vecchi, G.; Giannini, V.; Rivas, J.G. Shaping the fluorescent emission by lattice resonances in plasmonic crystals of nanoantennas. *Phys. Rev. Lett.* **2009**, *102*, 146807. [CrossRef]
29. Carron, K.T.; Fluhr, W.; Meier, M.; Wokaun, A.; Lehmann, H. Resonances of two-dimensional particle gratings in surface-enhanced Raman scattering. *JOSA B* **1986**, *3*, 430–440. [CrossRef]

30. Ye, J.; Wen, F.; Sobhani, H.; Lassiter, J.B.; Van Dorpe, P.; Nordlander, P.; Halas, N.J. Plasmonic nanoclusters: near field properties of the Fano resonance interrogated with SERS. *Nano Lett.* **2012**, *12*, 1660–1667. [CrossRef]
31. Kang, H.; Heo, C.J.; Jeon, H.C.; Lee, S.Y.; Yang, S.M. Durable plasmonic cap arrays on flexible substrate with real-time optical tunability for high-fidelity SERS devices. *ACS Appl. Mater. Interfaces* **2013**, *5*, 4569–4574. [CrossRef]
32. Félidj, N.; Aubard, J.; Lévi, G.; Krenn, J.R.; Salerno, M.; Schider, G.; Lamprecht, B.; Leitner, A.; Aussenegg, F. Controlling the optical response of regular arrays of gold particles for surface-enhanced Raman scattering. *Phys. Rev. B* **2002**, *65*, 075419. [CrossRef]
33. Moskovits, M. Surface-enhanced spectroscopy. *Rev. Mod. Phys.* **1985**, *57*, 783. [CrossRef]
34. Weitz, D.; Garoff, S.; Gersten, J.; Nitzan, A. The enhancement of Raman scattering, resonance Raman scattering, and fluorescence from molecules adsorbed on a rough silver surface. *J. Chem. Phys.* **1983**, *78*, 5324–5338. [CrossRef]
35. Le Ru, E.; Grand, J.; Felidj, N.; Aubard, J.; Levi, G.; Hohenau, A.; Krenn, J.; Blackie, E.; Etchegoin, P. Experimental verification of the SERS electromagnetic model beyond the |E|4 approximation: polarization effects. *J. Phys. Chem. C* **2008**, *112*, 8117–8121. [CrossRef]
36. Félidj, N.; Aubard, J.; Lévi, G.; Krenn, J.R.; Hohenau, A.; Schider, G.; Leitner, A.; Aussenegg, F.R. Optimized surface-enhanced Raman scattering on gold nanoparticle arrays. *Appl. Phys. Lett.* **2003**, *82*, 3095–3097. [CrossRef]
37. Gillibert, R.; Sarkar, M.; Bryche, J.F.; Yasukuni, R.; Moreau, J.; Besbes, M.; Barbillon, G.; Bartenlian, B.; Canva, M.; de La Chapelle, M.L. Directional surface enhanced Raman scattering on gold nano-gratings. *Nanotechnology* **2016**, *27*, 115202. [CrossRef]
38. Tinguely, J.C.; Sow, I.; Leiner, C.; Grand, J.; Hohenau, A.; Felidj, N.; Aubard, J.; Krenn, J.R. Gold nanoparticles for plasmonic biosensing: the role of metal crystallinity and nanoscale roughness. *BioNanoScience* **2011**, *1*, 128–135. [CrossRef]
39. Sow, I.; Grand, J.; Lévi, G.; Aubard, J.; Félidj, N.; Tinguely, J.C.; Hohenau, A.; Krenn, J. Revisiting surface-enhanced Raman scattering on realistic lithographic gold nanostripes. *J. Phys. Chem. C* **2013**, *117*, 25650–25658. [CrossRef] [PubMed]
40. Chu, Y.; Banaee, M.G.; Crozier, K.B. Double-resonance plasmon substrates for surface-enhanced Raman scattering with enhancement at excitation and stokes frequencies. *ACS Nano* **2010**, *4*, 2804–2810. [CrossRef] [PubMed]
41. Gotschy, W.; Vonmetz, K.; Leitner, A.; Aussenegg, F. Thin films by regular patterns of metal nanoparticles: tailoring the optical properties by nanodesign. *Appl. Phys. B* **1996**, *63*, 381–384. [CrossRef]
42. Hohenau, A.; Ditlbacher, H.; Lamprecht, B.; Krenn, J.R.; Leitner, A.; Aussenegg, F.R. Electron beam lithography, a helpful tool for nanooptics. *Microelectron. Eng.* **2006**, *83*, 1464–1467. [CrossRef]
43. Baida, F.I.; Belkhir, A. *Finite Difference Time Domain Method for Grating Structures*; Institut Fresnel, CNRS, Université d'Aix-Marseille: Marseille, France, 2012.
44. Berenger, J.P. A perfectly matched layer for the absorption of electromagnetic waves. *J. Comput. Phys.* **1994**, *114*, 185–200. [CrossRef]
45. Hamidi, M.; Baida, F.; Belkhir, A.; Lamrous, O. Implementation of the critical points model in a SFM-FDTD code working in oblique incidence. *J. Phys. Appl. Phys.* **2011**, *44*, 245101. [CrossRef]

Publisher's Note: MDPI stays neutral with regard to jurisdictional claims in published maps and institutional affiliations.

© 2020 by the authors. Licensee MDPI, Basel, Switzerland. This article is an open access article distributed under the terms and conditions of the Creative Commons Attribution (CC BY) license (http://creativecommons.org/licenses/by/4.0/).

Article

Hybrid Au/Si Disk-Shaped Nanoresonators on Gold Film for Amplified SERS Chemical Sensing

Grégory Barbillon [1],*, Andrey Ivanov [2] and Andrey K. Sarychev [2]

1 EPF-Ecole d'Ingénieurs, 3 bis rue Lakanal, 92330 Sceaux, France
2 Institute for Theoretical and Applied Electrodynamics, Russian Academy of Sciences, 125412 Moscow, Russia; av.ivanov@physics.msu.ru (A.I.); sarychev_andrey@yahoo.com (A.K.S.)
* Correspondence: gregory.barbillon@epf.fr

Received: 15 October 2019; Accepted: 5 November 2019; Published: 8 November 2019

Abstract: We present here the amplification of the surface-enhanced Raman scattering (SERS) signal of nanodisks on a gold film for SERS sensing of small molecules (thiophenol) with an excellent sensitivity. The enhancement is achieved by adding a silicon underlayer for the composition of the nanodisks. We experimentally investigated the sensitivity of the suggested Au/Si disk-shaped nanoresonators for chemical sensing by SERS. We achieved values of enhancement factors of $5 \times 10^7 - 6 \times 10^7$ for thiophenol sensing. Moreover, we remarked that the enhancement factor (EF) values reached experimentally behave qualitatively as those evaluated with the E^4 model.

Keywords: SERS; sensors; plasmonics; gold; silicon

1. Introduction

Surface-enhanced Raman scattering (SERS) is often employed as a fast technique of analysis owing to a high sensitivity for sensing of different types of molecules [1–3]. In SERS, the dominant contribution is the electromagnetic mechanism [2,4] allowing the obtaining of very high enhancement factors (EF). This EF for SERS is evaluated as the fourth power of the intensity of the local electric field [5,6]. Thus, the design of nanostructures to achieve high enhancement factors in the research domain of SERS is a very important point in order to increase the sensitivity of the biological and chemical sensing. Modern micro/nanofabrication tools such as focused ion-beam lithography [7], electron-beam lithography [8–11], X-ray, deep UV, UV, and interference lithographies [12–16] favor the numerous designs of SERS substrates with an accuracy control over the shape and spatial distribution of nanostructures. Furthermore, some low cost techniques of fabrication as nanoimprint lithography (NIL) [17,18] and nanosphere lithography (NSL) [19–22] may enable the realization of these SERS substrates. A large number of nanostructures such as nanodisks, nanoholes, nanodimers have been tested and provided high EFs for SERS [23–25]. The majority of these designs are focused on the control of the resonances of localized surface plasmons (LSPR) for optimizing the SERS enhancement [26,27]. In addition, a significant improvement of strong electric field zones around the metallic nanostructures can be observed by adding a metallic film under the plasmonic nanostructures. This enhancement is obtained thanks to the coupling between the nanostructures (antennas) via surface plasmon polaritons on the Au film [28,29] or localized surface plasmon hybridization with the image modes in a plasmonic substrate [30,31]. Thus, this supplementary enhancement can be exploited to amplify the SERS effect [21,32–34]. Another pathway for realizing significant EFs is to employ Si nanowires (SiNW) or nanopillars (SiNP) coupled to metallic nanoparticles or covered by a metallic layer allowing thus the obtaining of a better detection limit [35–43]. Moreover, fabrication techniques of large-surface may allow the realization of disordered Si nanowires. Another possibility is to realize tip-shaped Si metasurface on which metallic nanoparticles are deposited [44,45]. In addition, the Moskovits group has demonstrated that the substantial input to the SERS enhancement, for silicon/silica/metal nanogratings, is a non-local (plasmonic) effect of grating depending mainly on the grating parameters until the metal conductivity is not sufficient [46].

The main goal of this paper is to improve the SERS effect of gold nanodisks on a gold film by the simple addition of a silicon layer for the composition of the nanodisks (between the gold film and the gold layer of nanodisks). In such hybrid nanostructures, we use the second dipole resonance for enhancing the SERS signal compared to our previous works [11,41]. Moreover, these hybrid Au/Si disk-shaped nanoresonators on the gold film have been tested as chemical sensors by using solutions of thiophenol, which are small chemical molecules (thickness of a thiophenol monolayer is around 0.6 nm [47]). Besides, this additional layer of silicon has already allowed the improvement of the fluorescence signal enhanced by the surface for biosensing applications and enhanced single-molecule detection [48,49].

2. Experimental Details

2.1. Fabrication of Hybrid Au/Si Nanodisks

The hybrid Au/Si nanodisk (ND) fabrication is divided into several steps: (i) evaporation of a gold layer under vacuum by electron-beam (EBE) on Si substrate covered of a Ti adhesion layer for Au (2 nm), (ii) electron beam lithography, (iii) deposition of Si and Au layers, and (iv) lift-off in acetone. Firstly, a gold layer (thickness of 40 nm) was evaporated on Si substrate by EBE under normal incidence. Next, we deposited a PMMA layer (polymethylmethacrylate A2: thickness of 90 nm) by spin-coating on gold film. Then, several 300×300 μm^2 arrays of nanodisks were realized by electron beam lithography (NanoBeam). Next, the sample was immersed in a development solution of 1:3 methylisobutylketone/isopropanol (MIBK/IPA) in order to reveal nanodisks. The next step consisted of an evaporation of a 20-nm silicon layer following by a second evaporation of a 20-nm gold layer both realized by EBE. Finally, a lift-off process in acetone was employed in order to obtain the hybrid Au/Si nanodisks on the gold film (see Figure 1). The evaporation rates used in this fabrication are 0.05 nm/s, 0.1 nm/s and 0.3 nm/s for Ti, Si and Au layers, respectively. In addition, geometrical parameters of hybrid nanodisks are a diameter of 130 nm (D), a period of 300 nm (P), and a total height of 40 nm (20 nm of Si + 20 nm of Au). In addition, we chose a Ti adhesion layer of 2 nm in order to have a good compromise between the adhesion properties and the electric field enhancement. Indeed, in the spectral range of our study, Ti is a material less absorptive than Cr, which is another material widely used as adhesion layer, and consequently, Ti reduces the electric field enhancement less than Cr [50].

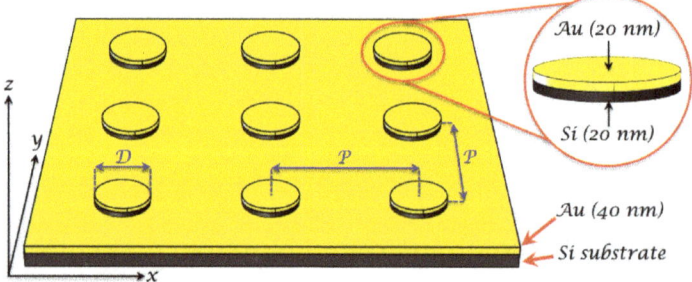

Figure 1. Scheme of the Au/Si disk-shaped nanoresonator array on gold film. D and P correspond to the nanodisk (ND) diameter and the period between the nanodisks, respectively. P is identical along x-axis and y-axis. In the red zoom are indicated the thicknesses of Si and Au layers constituting the bilayer of a hybrid nanodisk. An adhesion layer of Ti (2 nm) is used between Si substrate and gold film.

2.2. Thiophenol Functionalization of the Hybrid Au/Si Nanodisks

For investigating the SERS performances of our hybrid NDs on a gold film, thiophenol molecules were used as probe molecules for their efficient grafting on metallic surfaces. The functionalization constituted of four steps: (i) realization of a thiophenol solution in ethanol (1 mM); (ii) dipping for 24 h the SERS substrate in the thiophenol solution freshly prepared (obtaining of a thiophenol monolayer on the gold parts); (iii) washing the SERS sample by using ethanol and (iv) drying it by using compressed nitrogen. For our Raman experiments in a solution which serves as reference, a highly concentrated solution of thiophenol in ethanol (1 M) was used because the Raman cross-sections of thiophenol in solution are very low.

2.3. Raman Spectroscopy of Hybrid Nanodisks on Gold Film

We employed a Labram spectrophotometer (Horiba Scientific) with a spectral resolution of 1 cm^{-1}. For all the SERS and Raman (reference) measurements, we have set the excitation wavelength at 785 nm (λ_{exc}) and the acquisition time at 10 s. Concerning to the SERS measurements, a microscope objective (×100, N.A. = 0.9) was used in order to concentrate the laser beam on the sample. Then, SERS signal coming from the hybrid samples was detected by this same objective configured in a backscattering setup. The laser power for the excitation wavelength of 785 nm was 3 mW. Besides, for Raman measurements serving as reference, the same excitation wavelength and a macro-objective of which the focal length is 40 mm (N.A. = 0.18) were employed. All recorded spectra have been divided by the acquisition time and the laser power for comparison purposes.

2.4. Plasmon Resonances in Hybrid Nanodisks on Gold Film

To get inside in the plasmon resonances that are responsible for the observed SERS, we consider the plasmon resonator composed of two gold disks with a silicon layer between them (see Figure 1). Suppose, for the beginning that metal plates are optically thick, i.e., $h|m|k \gg 1$, where h is the metal plate thickness, $m = \sqrt{-\varepsilon_m}$ is the metal "refractive index", and the wave-vector is $k = 2\pi/\lambda$. We also suppose that the thickness d of the silicon layer is much smaller than the radius a of the resonator. Then, the distortion of the EM field near the outer boundary ($r \leq a$) of the resonator can be neglected. The cylindrical coordinates $\{r, \varphi, z\}$ are used below so that the z-axis coincides with the axis of the resonator, axes origin is in the center of the resonator. The plasmon electromagnetic field in the resonator in the dipole mode can be found from the vector potential **A** that has z-component only.

$$A_z^{(1)} = \exp[q_2(z+d/2)] J_1(qr) \sin(\varphi), \quad -h-d/2 < z < -d/2; \quad (1)$$

$$A_z^{(2)} = \frac{\cosh(q_1 z)}{\cosh(dq_1/2)} J_1(qr) \sin(\varphi), \quad -d/2 < z < d/2; \quad (2)$$

$$A_z^{(3)} = \exp[-q_2(z-d/2)] J_1(qr) \sin(\varphi), \quad z > d/2, \quad (3)$$

where the silicon layer with refraction index n is placed in the gap $-d/2 < z < d/2$ between two metal plates, $J_1(qr)$ is the Bessel function of the first order, $q_1 = \sqrt{q^2 - k^2 n^2}$ and $q_2 = \sqrt{q^2 + k^2 m^2}$ are the wave-vectors. The vector potentials thus defined are the solutions of the wave equations, namely, $(\Delta - (mk)^2) A_z^{(1,3)} = 0$ and $(\Delta + (nk)^2) A_z^{(2)} = 0$, where the symbol Δ stands for the Laplace operator. The electric and magnetic fields in the resonator are given by the Maxwell equations $\mathbf{H}^{(j)} = \text{curl } \mathbf{A}^{(j)}$, $\mathbf{E}^{(j)} = i\text{curl } \mathbf{H}^{(j)} / [k\varepsilon^{(j)}]$, where $j = 1, 2, 3$ correspond to the upper metal plate, silicon layer, and lower metal plate correspondingly, so that $\varepsilon^{(1,3)} = \varepsilon_m \equiv -m^2$ and $\varepsilon^{(2)} = n^2$. Thus, at the middle plane $z = 0$, the electric field has z-component only, which equals to:

$$E_z = -i \frac{q^2 J_1(qr)}{kn^2 \cosh(dq_1/2)} \cos(\varphi), \quad (4)$$

while z-component of the magnetic field equals to zero everywhere. The magnetic field in the gold and silicon, obtained from vector potentials in Equations (1)–(3), has components H_x and H_y only in contrast to the electric field that has all three components. The magnetic field of the plasmon is perpendicular to the axis of the cylinder (z-axis) and plasmon can be called HT plasmon. Matching the fields at the metal-dielectric interfaces $z = \pm d/2$, we obtain the dispersion equation for the wave-vector q of the HT plasmon excited in the disc resonator:

$$n^2 q_1 = m^2 q_2 \tanh\left(dq_2/2\right). \tag{5}$$

In a thin resonator $d|n|k \ll 1$, Equation (5) has the simple analytical solution as follows:

$$q = q^{(s)} = 2 \operatorname{arctanh}\left(n^2/m^2\right)/d, \tag{6}$$

Therefore, the plasmon is effectively excited when metal refractive index m is larger in absolute value than the silicon refractive index $|m| > |n|$. The absolute value of gold permittivity is large in the optical spectral range, however, silicon permittivity $|n|^2 \simeq 15$ is also large (see [51–53]). Then, the condition $|m| > |n|$ is violated in the gold–silicon resonator for wavelength $\lambda < 600$ nm. For smaller wavelengths, the antisymmetrical plasmon can be excited that vector potential is still given by Equations (1)–(3), where $\cosh(\ldots)$ in Equation (2) should be replaced by $\sinh(\ldots)$ and $A_z^{(3)}$ is taken with opposite sign. The dispersion equation for the antisymmetrical plasmon takes form:

$$n^2 q_1 = m^2 q_2 \coth\left(dq_2/2\right). \tag{7}$$

In the thin resonator $d|n|k \ll 1$, Equation (7) has the simple analytical solution as follows:

$$q = q^{(a)} = 2 \operatorname{arctanh}\left(m^2/n^2\right)/2, \tag{8}$$

Therefore, the antisymmetrical plasmon is effectively excited when metal refractive index m is smaller in absolute value than the silicon refractive index $|m| < |n|$. All plasmons discussed above could be excited simultaneously when the upper gold plate has finite thickness and radiation from the resonator cannot be neglected. Then, the energy absorption as a function of λ has set of maxima, and reflectance $R(\lambda)$ has many peculiarities as displayed in Figure 2.

Suppose that the gold–silicon–gold disk resonator is illuminated from the top. The lower metal plate is still considered as optically thick. The vector potential in the resonator can be considered as a superposition of the symmetric and antisymmetric plasmons

$$A_z^{(1)} = [a_1 \exp(q_2 z) + a_2 \exp(-q_2 z)] J_1(qr) \sin(\varphi), \quad -h - d/2 < z < -d/2; \tag{9}$$
$$A_z^{(2)} = [b_1 \exp(q_1 z) + b_2 \exp(-q_1 z)] J_1(qr) \sin(\varphi), \quad -d/2 < z < d/2; \tag{10}$$
$$A_z^{(3)} = c_2 \exp(-q_2 z) J_1(qr) \sin(\varphi), \quad z > d/2; \tag{11}$$

where coefficients a_1, a_2, b_1, b_2, c_2 are obtained by matching magnetic and electric fields at the interfaces between gold and silicon at $z = \pm d/2$. We apply the boundary condition $J_0(q_p a) = 0$ at the lateral boundary and found set of the harmonics p; $J_0(x)$ is the Bessel function of zero order. The incident and reflected electromagnetic waves are expanded in series of these harmonics and match the EM field in the resonator at the top of the resonator $z = -h - d/2$. When EM field in the resonator is known, we can calculate EM wave, which is radiated by the periodic system of the resonators shown in Figure 1. This wave is added to the wave reflected by the bare gold film on the top of the silicon substrate. Thus, obtained reflectance $R(\lambda)$ is shown in Figure 2 together with results of the computer simulation.

We performed computer simulations of the periodic array of the disk resonators in the COMSOL environment. The incident light was normal to the film plane. The Maxwell equations were solved by using the finite element method (FEM). The geometrical parameters of this model are: the top nanodisk has diameter of $D = 2a = 130$ nm, thicknesses of gold and silicon disks are 20 nm. The underneath gold film had a thickness of 40 nm. The nanodisks were organized in the square lattice with a periodicity of $P = 300$ nm (see Figures 1 and 3). There was a qualitative agreement between computer simulations and the discussed simple analytical model.

Dips in the reflectance $R(\lambda)$, which are well seen in Figure 2, correspond to the various plasmon resonances. Minima at $\lambda \simeq 1400$ nm and $\lambda \simeq 800$ nm correspond to the first and second dipole resonances, respectively. Minima at shorter wavelengths are due to the higher symmetric as well as antisymmetric plasmon modes. Since the absolute value of the gold permittivity is on the order of the silicon permittivity, the resonances could be rather wide. The simulation results obtained for the electric field in the disk resonator are shown in Figure 4, where a dipole mode can be seen. The field spread over the entire resonator. This form of the resonance field is different from the field distribution obtained for a similar system in [54], where a 5-nm SiO_2 layer was between the plates. We speculate the permittivity of the silica is well much lower than the absolute value of the gold permittivity and the EM field is confined in the resonator. The electric field, presented in Figure 4, is enhanced at the upper rim of the cylinder resonator. The field distribution is similar to the field calculated in [55].

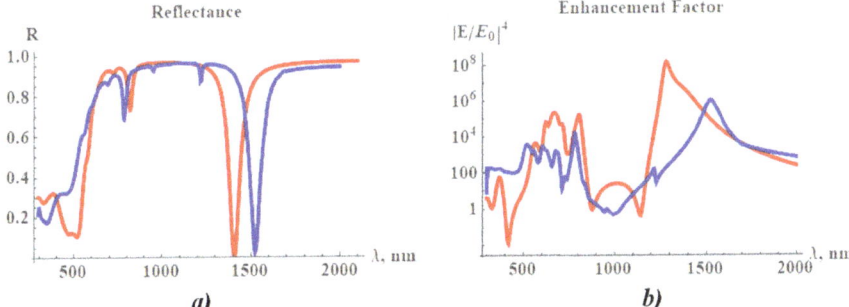

Figure 2. Results of analytical model (red) and COMSOL computer simulations (blue) of the system shown in Figure 1, disk diameter $D = 2a = 130$ nm, period of square lattice $P = 300$ nm, thicknesses of the upper gold plate and silicon interlayer are 20 nm, gold film has thickness of 40 nm. (**a**) Reflectance from surface-enhanced Raman scattering (SERS) substrates and (**b**) electric field enhancement factor (EF) averaged over the lateral side of the plasmon resonator EF = $\langle |E/E_0|^4 \rangle$, where E_0 corresponds to the amplitude of the incident EM wave.

Enhancement factor of the electric field averaged over lateral side of the resonator is shown in Figure 2. The enhancement $|E/E_0|^4$ achieves $\sim 10^6$ at $\lambda \simeq 800$ nm resonance and it takes even large values $\sim 10^8$ at $\lambda \simeq 1400$ nm. Reflectance $R(\lambda)$ has wide minimum at $\lambda > 500$ nm, however, the silicon as well as gold have large ohmic loss for $\lambda > 500$ nm, and electric field is not much enhanced in this spectral band as it is seen in Figure 2b.

Since the Au film thickness is larger compared to the skin-depth (~ 30 nm) and silicon plate is optically thick (opaque), the extinction (or absorbance) equals to $A = 1 - R$, where the reflectance R is discussed above.

3. Results and Discussion

3.1. Fabrication and Functionalization of the Hybrid Disk-Shaped Nanoresonators on Gold Film

Firstly, the hybrid Au/Si disk-shaped nanoresonators on gold film have been realized by using the process of Section 2.1. A SEM picture of the obtained Au/Si nanodisks is displayed in Figure 3. The diameter, periodicity and height obtained for the hybrid NDs are 130 nm, 300 nm and 40 nm, respectively.

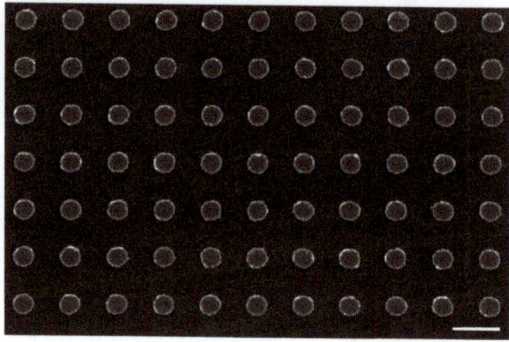

Figure 3. SEM picture of a hybrid Au/Si nanodisk array on gold film (scale bar = 300 nm). The nanodisk dimensions are 130 nm of diameter, 40 nm of total height, and 300 nm of periodicity.

The next step was the functionalization of thiophenol molecules on hybrid Au/Si ND array by employing the protocol of the Section 2.2. Raman measurements were realized immediately after this functionalization step. SERS spectra of thiophenol on Au/Si nanodisks arrays on gold film obtained for the excitation wavelength of 785 nm are displayed in Figure 4a. From these spectra, Raman peaks of thiophenol molecules were observed (see refs [56,57]) of which those at 1000 cm^{-1} coincided with the association of certain modes: C–H out-of-plane bending and ring out-of-plane deformation (named: $\gamma(CH)$ and $r-o-d$); at 1025 cm^{-1} coinciding with the association of other modes: ring in-plane deformation and C–C symmetric stretching (named: $r-i-d$ and $\nu(CC)$); at 1075 cm^{-1} coinciding also with the association of other modes: C–C symmetric stretching and C-S stretching (named: $\nu(CC)$ and $\nu(CS)$, respectively), and at 1575 cm^{-1} coinciding with the C–C symmetric stretching mode (named: $\nu(CC)$). Besides, a couple of peaks located in the domain of 900–980 cm^{-1} is present and corresponding to multiphonon peaks of Si [58,59].

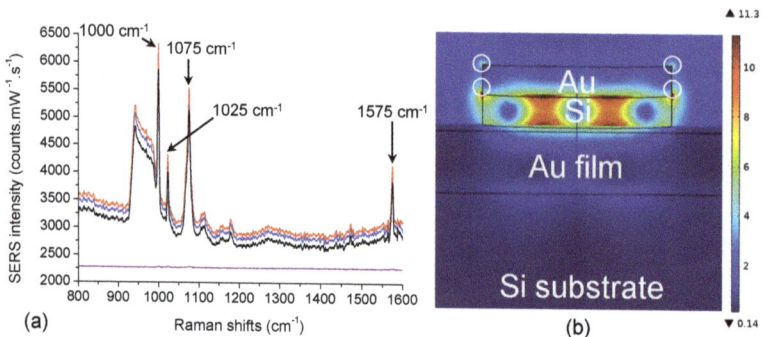

Figure 4. (a) SERS spectra of thiophenol realized on 3 distinct SERS substrates among 10 for the excitation wavelength of 785 nm. In purple is represented the SERS spectrum of thiophenol (1 mM) obtained on a 40-nm gold film at the same excitation wavelength (an offset is applied to the purple spectrum to see all the SERS spectra). (b) Electric field mapping $|E/E_0|$ of a hybrid Au/Si nanodisk on gold film for an excitation wavelength of 785 nm (cross-sectional view). White circles correspond to the strong electric field zones accessible for thiophenol molecules.

3.2. Sensitivity of the Hybrid Disk-Shaped Nanoresonators and Reproducibility of SERS Signal

For examining the detection sensitivity of hybrid Au/Si ND arrays, the EF is evaluated for the 4 previous Raman peaks by the following formula:

$$EF = \frac{I_{SERS}}{I_{Raman}} \times \frac{N_{Raman}}{N_{SERS}} \quad (12)$$

where I_{SERS}, I_{Raman} represent the SERS and Raman intensities, respectively (see Table 1). N_{SERS}, N_{Raman} are the numbers of thiophenol molecules for SERS and reference Raman experiments, respectively. N_{SERS} is determined by this formula:

$$N_{SERS} = N_A \times S_{illuminated} \times \sigma_{Surf} \quad (13)$$

where N_A is the Avogadro's number (mol^{-1}), $S_{illuminated}$ corresponds here to the lateral surface (gold part) of one nanodisk (ND surface: S = 8.2 × 10^3 nm^2) which is multiplied by the number of nanodisks (~12) illuminated in the laser spot of which the size is about ~1 µm^2 for λ_{exc} = 785 nm. σ_{Surf} represents the surface coverage of thiophenol (here σ_{Surf} = 0.544 nmol/cm^2) [60,61]. Thus, thiophenol molecules of interest were grafted on lateral gold parts of hybrid nanodisks, and the number of excited molecules N_{SERS} is 3.22 × 10^5 for the excitation wavelength of 785 nm. Furthermore, no SERS signal is recorded from the smooth gold film (see Figure 4a, and as also observed in our previous works [10,11]). Moreover, we observed from the electric field mapping (see Figure 4b) that the effective SERS signals (strong electric field zones accessible for thiophenol molecules, see the white circles on Figure 4b) are localized at the interface between silicon and gold layers around the hybrid nanodisk, and also at the top of the lateral surface of the gold layer. Thus, from these observations, we speculate that the equivalent surface of interest for evaluation of EF is the lateral surface of the gold part of the hybrid nanodisk. For the Raman measurements serving as reference, the number N_{Raman} is 4.24 × 10^{11} for the excitation wavelength of 785 nm. This value of N_{Raman} is obtained by this expression:

$$N_{Raman} = N_A \times C \times V_{sca}, \quad (14)$$

where C and V_{sca} correspond to the concentration used for thiophenol molecules (1 M), and the scattering volume, respectively. This latter is determined by this formula: V_{sca} = A × H, where A is the scattering area corresponding to the disk area with a diameter of 5.3 µm at 785 nm, and H (scattering height, see Refs [62,63]) of approximately 32 µm for λ_{exc} = 785 nm. Thus, V_{sca} is equal to 704 µm^3~0.704 pL.

Table 1. For λ_{exc} = 785 nm, and the four Raman shifts (RS) of thiophenol, λ_{Raman} coinciding with RS, the intensities I_{Raman} and I_{SERS}, relative standard deviations (RSDs) coinciding with I_{SERS}, EF obtained with Equation (1) and EF values (in arbitrary unit, see Equation (5)) calculated with the E^4 model are tabulated.

Number	RS (cm^{-1})	$\lambda_{Raman}(nm)$	I_{Raman}	I_{SERS}	RSD (%)	EF	EF (a.u.)
1	1000	852	79	3533	4.9	5.9 × 10^7	0.0198
2	1025	854	32	1402	4.1	5.8 × 10^7	0.0192
3	1075	857	55	2350	4.7	5.6 × 10^7	0.0186
4	1575	896	18	724	5.0	5.3 × 10^7	0.0170

From the results summarized in Table 1, EF values were found in the range of 5 × 10^7–6 × 10^7. Likewise, several groups showed good EF with similar SERS substrates composed of regular metallic nanostructures on a metallic film, such as gold nanodisks on a gold film (EF ~ 10^3–10^4 in reference [8], and EF ~ 10^6–10^7 in reference [11]), and 3D donut-like gold nanorings on a gold film (EF = 3.84 × 10^7 in reference [64]). By comparing them, we remarked that our hybrid disk-shaped nanoresonators achieved higher EFs. In addition, in order to assess the substrate-to-substrate reproducibility for the SERS signal, the relative standard deviation (RSD) is evaluated for each Raman peak studied here. Each RSD value is obtained from the measurements of the SERS signal on 10 distinct substrates on which this SERS signal was recorded on 4 arrays of hybrid nanodisks (300 × 300 µm^2) under same experimental conditions. Thus, the RSD values were obtained from 40 SERS spectra (see 3 examples in Figure 4a). Finally, a very fine substrate-to-substrate reproducibility for the SERS signal is reached for all the Raman peaks studied here (RSD ⩽ 5%, see Table 1).

3.3. Spectral Analysis

The extinction spectrum of the hybrid disk-shaped nanoresonators has been calculated by using numerical simulations (see Figure 5) in order to qualitatively compare the behavior of the experimental EF values with this of EF values obtained with the E^4 model. The wavelengths of different resonances of hybrid NDs and the excitation and Raman wavelengths can be compared. The following expression enabled us determining the Raman scattering wavelength (λ_{Raman}):

$$\Delta \omega = 10^7 \left(\frac{1}{\lambda_{exc}} - \frac{1}{\lambda_{Raman}} \right) \tag{15}$$

where $\Delta \omega$ (cm^{-1}), λ_{exc} (nm) and λ_{Raman} (nm) are the Raman shift, the excitation and Raman scattering wavelengths, respectively (see Table 1). In this E^4 model, EF is presumed to be comparable to the extinction intensities (Q_e) at λ_{exc} and λ_{Raman} [65] as follows:

$$EF \sim Q_e(\lambda_{exc}) \times Q_e(\lambda_{Raman}). \tag{16}$$

From Figure 5 and Table 1, EF$_1$ corresponds to the largest value that we observed, and EFs decreased when λ_{Raman} increased, i.e., $Q_e(\lambda_{Raman})$ decreased with λ_{Raman}. The different EF values (from EF$_1$ to EF$_4$) match to EFs concerning to the couples (λ_{exc}, λ_{Raman1}), (λ_{exc}, λ_{Raman2}), (λ_{exc}, λ_{Raman3}) and (λ_{exc}, λ_{Raman4}), respectively. Thus, we observed that the EFs achieved experimentally (see Table 1) behave qualitatively as those evaluated with the E^4 model.

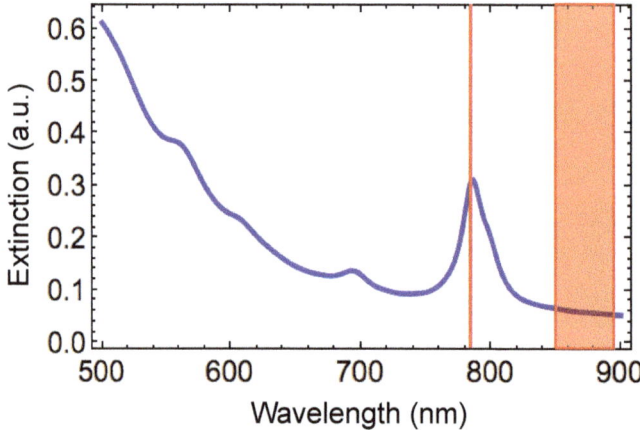

Figure 5. Calculated extinction spectrum of the hybrid Au/Si disk-shaped nanoresonators. The red line matches to the excitation wavelength of 785 nm. The full red rectangle represents all the Raman wavelengths (λ_{Raman}) corresponding to the associated Raman shifts (from 1000 to 1575 cm^{-1}, see Table 1).

4. Conclusions

We showed the amplification of the SERS signal of nanodisks on a gold film by a simple addition of a silicon layer for the composition of the nanodisks. The sensitivity of these hybrid SERS substrates has been studied and compared to the results in literature obtained for regular gold nanostructures on a gold film. The EF values reached with the suggested SERS substrates ($5 \times 10^7 <$ EF $< 6 \times 10^7$) are larger than EFs cited above. We remarked that the experimental EF values have the same behavior as those obtained with the E^4 model by using a generic analytical approach and numerical simulations. Our hybrid Au/Si disk-shaped nanoresonators on gold film can be optimized in order to obtain even higher enhancement factors. The obtained SERS substrates offer the possibility of being incorporated on a lab-on-chip for a label-free sensor of biochemical species in the nearest future.

Author Contributions: G.B. conceived the research, performed the experiments, wrote the whole paper, prepared the original draft, edited the draft; A.I. performed the simulations, wrote the Section 2.4, edited the draft; A.K.S. suggested analytical theory, performed the simulations, wrote the Section 2.4, edited the draft.

Funding: This research received external funding from the Russian Foundation for Basic Research (Grant No. 17-08-01448 and 18-58-00048), Russian Science Foundation (grant No. 16-14-00209), the Presidium of RAS (Program 22).

Conflicts of Interest: The authors declare no conflict of interest.

References

1. Sharma, B.; Frontiera, R.R.; Henry, A.-I.; Ringe, E.; Van Duyne, R.P. SERS: Materials, applications, and the future. *Mater. Today* **2012**, *15*, 16–25. [CrossRef]
2. Ding, S.-Y.; Yi, J.; Li, J.-F.; Ren, B.; Wu, D.-Y.; Panneerselvam, R.; Tian, Z.-Q. Nanostructure-based plasmon-enhanced Raman spectroscopy for surface analysis of materials. *Nat. Rev. Mater.* **2016**, *1*, 16021. [CrossRef]
3. Barbillon, G. *Nanoplasmonics-Fundamentals and Applications*; InTech: Rijeka, Croatia, 2017; pp. 3–481.
4. Wustholz, K.L.; Henri, A.-I.; McMahon, J.M.; Freeman, R.G.; Valley, N.; Piotti, M.E.; Natan, M.J.; Schatz, G.C.; Van Duyne, R.P. Structure–Activity Relationships in Gold Nanoparticle Dimers and Trimers for Surface-Enhanced Raman Spectroscopy. *J. Am. Chem. Soc.* **2010**, *132*, 10903–10910. [CrossRef]
5. Itoh, T.; Yamamoto, Y.S.; Ozaki, Y. Plasmon-enhanced spectroscopy of absorption and spontaneous emissions explained using cavity quantum optics. *Chem. Soc. Rev.* **2017**, *49*, 3904–3921. [CrossRef] [PubMed]
6. Ding, S.-Y.; You, E.-M.; Tian, Z.-Q.; Moskovits, M. Electromagnetic theories of surface-enhanced Raman spectroscopy. *Chem. Soc. Rev.* **2017**, *46*, 4042–4076. [CrossRef] [PubMed]
7. Henzie, J.; Lee, J.; Lee, M.H.; Hasan, W.; Odom, T.W. Nanofabrication of Plasmonic Structures. *Annu. Rev. Phys. Chem.* **2009**, *60*, 147–165. [CrossRef] [PubMed]
8. Yu, Q.; Guan, P.; Qin, D.; Golden, G.; Wallace, P.M. Inverted size-dependence of surface-enhanced Raman scattering on gold nanohole and nanodisk arrays. *Nano Lett.* **2008**, *8*, 1923–1928. [CrossRef]
9. Faure, A.C.; Barbillon, G.; Ou, M.; Ledoux, G.; Tillement, O.; Roux, S.; Fabregue, D.; Descamps, A.; Bijeon, J.-L.; Marquette, C.A.; et al. Core/shell nanoparticles for multiple biological detection with enhanced sensitivity and kinetics. *Nanotechnology* **2008**, *19*, 485103. [CrossRef]
10. Bryche, J.-F.; Gillibert, R.; Barbillon, G.; Sarkar, M.; Coutrot, A.-L.; Hamouda, F.; Aassime, A.; Moreau, J.; Lamy de la Chapelle, M.; et al. Density effect of gold nanodisks on the SERS intensity for a highly sensitive detection of chemical molecules. *J. Mater. Sci.* **2015**, *50*, 6601–6607. [CrossRef]
11. Bryche, J.-F.; Gillibert, R.; Barbillon, G.; Gogol, P.; Moreau, J.; Lamy de la Chapelle, M.; Bartenlian, B.; Canva, M. Plasmonic enhancement by a continuous gold underlayer: Application to SERS sensing. *Plasmonics* **2016**, *11*, 601–608. [CrossRef]
12. Zhang, P.; Yang, S.; Wang, L.; Zhao, J.; Zhu, Z.; Liu, B.; Zhong, J.; Sun, X. Large-scale uniform Au nanodisk arrays fabricated via X-ray interference lithography for reproducible and sensitive SERS substrate. *Nanotechnology* **2014**, *25*, 245301. [CrossRef] [PubMed]
13. Barbillon, G.; Bijeon, J.-L.; Lérondel, G.; Plain, J.; Royer, P. Detection of chemical molecules with integrated plasmonic glass nanotips. *Surf. Sci.* **2008**, *602*, L119–L122. [CrossRef]

14. Dhawan, A.; Duval, A.; Nakkach, M.; Barbillon, G.; Moreau, J.; Canva, M.; Vo-Dinh, T. Deep UV nano-microstructuring of substrates for surface plasmon resonance imaging. *Nanotechnology* **2011**, *22*, 165301. [CrossRef] [PubMed]
15. Guisbert Quilis, N.; Lequeux, M.; Venugopalan, P.; Khan, I.; Knoll, W.; Boujday, S.; Lamy de la Chapelle, M.; Dostalek, J. Tunable laser interference lithography preparation of plasmonic nanoparticle arrays tailored for SERS. *Nanoscale* **2018**, *10*, 10268. [CrossRef] [PubMed]
16. Hwang, J.S.; Yang, M. Sensitive and Reproducible Gold SERS Sensor Based on Interference Lithography and Electrophoretic Deposition. *Sensors* **2018**, *18*, 4076. [CrossRef] [PubMed]
17. Ding, T.; Sigle, D.O.; Herrmann, L.O.; Wolverson, D.; Baumberg, J.J. Nanoimprint lithography of Al Nanovoids for Deep-UV SERS. *ACS Appl. Mater. Interfaces* **2014**, *6*, 17358–17363. [CrossRef]
18. Cottat, M.; Lidgi-Guigui, N.; Tijunelyte, I.; Barbillon, G.; Hamouda, F.; Gogol, P.; Aassime, A.; Lourtioz, J.-M.; Bartenlian, B.; de la Lamy Chapelle, M. Soft UV nanoimprint lithography-designed highly sensitive substrates for SERS detection. *Nanoscale Res. Lett.* **2014**, *9*, 623. [CrossRef]
19. Masson, J.F.; Gibson, K.F.; Provencher-Girard, A. Surface-enhanced Raman spectroscopy amplification with film over etched nanospheres. *J. Phys. Chem. C* **2010**, *114*, 22406–22412. [CrossRef]
20. Bechelany, M.; Brodard, P.; Elias, J.; Brioude, A.; Michler, J.; Philippe, L. Simple Synthetic Route for SERS-Active Gold Nanoparticles Substrate with Controlled Shape and Organization. *Langmuir* **2010**, *26*, 14364–14371. [CrossRef]
21. Bryche, J.-F.; Tsigara, A.; Bélier, B.; Lamy de la Chapelle, M.; Canva, M.; Bartenlian, B.; Barbillon, G. Surface enhanced Raman scattering improvement of gold triangular nanoprisms by a gold reflective underlayer for chemical sensing. *Sens. Actuators B* **2016**, *228*, 31–35. [CrossRef]
22. Barbillon, G.; Noblet, T.; Busson, B.; Tadjeddine, A.; Humbert, C. Localised detection of thiophenol with gold nanotriangles highly structured as honeycombs by nonlinear sum frequency generation spectroscopy. *J. Mater. Sci.* **2018**, *53*, 4554–4562. [CrossRef]
23. Brolo, A.G.; Arctander, E.; Gordon, R.; Leathem, B.; Kavanagh, K.L. Nanohole-Enhanced Raman Scattering. *Nano Lett.* **2004**, *4*, 2015–2018. [CrossRef]
24. Suh, J.Y.; Odom, T.W. Nonlinear properties of nanoscale antennas. *Nano Today* **2013**, *8*, 469–479. [CrossRef]
25. Lim, D.-K.; Jeon, K.-S.; Kim, H.M.; Nam, J.-M.; Suh, Y.D. Nanogop-engineerable Raman-active nanodumbbells for single-molecule detection. *Nat. Mater.* **2010**, *9*, 60–67. [CrossRef] [PubMed]
26. Félidj, N.; Aubard, J.; Lévi, G.; Krenn, J.R.; Hohenau, A.; Schider, G.; Leitner, A.; Aussenegg, F.R. Optimized surface-enhanced Raman scattering on gold nanoparticle arrays. *Appl. Phys. Lett.* **2003**, *82*, 3095–3097. [CrossRef]
27. Guillot, N.; Shen, H.; Frémaux, B.; Péron, O.; Rinnert, E.; Toury, T.; Lamy de la Chapelle, M. Surface enhanced Raman scattering optimization of gold nanocylinder arrays: Influence of the localized surface plasmon resonance and excitation wavelength. *Appl. Phys. Lett.* **2010**, *97*, 023113. [CrossRef]
28. Li, Z.; Butun, S.; Aydin, K. Ultranarrow Band Absorbers Based on Surface Lattice Resonances in Nanostructured Metal Surfaces. *ACS Nano* **2014**, *8*, 8242–8248. [CrossRef]
29. Sarkar, M.; Besbes, M.; Moreau, J.; Bryche, J.-F.; Olivéro, A.; Barbillon, G.; Coutrot, A.-L.; Bartenlian, B.; Canva, M. Hybrid Plasmonic Mode by Resonant Coupling of Localized Plasmons to Propagating Plasmons in a Kretschmann Configuration. *ACS Photonics* **2015**, *2*, 237–245. [CrossRef]
30. Sobhani, A.; Manjavacas, A.; Cao, Y.; McClain, M.J.; Javier Garcia de Abajo, F.; Nordlander, P.; Halas, N.J. Pronounced Linewidth Narrowing of an Aluminum Nanoparticle Plasmon Resonance by Interaction with an Aluminum Metallic Film. *Nano Lett.* **2015**, *15*, 6946–6951. [CrossRef]
31. Yue, W.; Wang, Z.; Whittaker, J.; Lopez-Royo, F.; Yang, Y.; Zayats, A.V. Amplification of surface-enhanced Raman scattering due to substrate-mediated localized surface plasmons in gold nanodimers. *J. Mater. Chem. C* **2017**, *5*, 4075–4084. [CrossRef]
32. Zhou, Q.; Liu, Y.; He, Y.; Zhang, Z.; Zhao, Y. The effect of underlayer thin films on the surface-enhanced Raman scattering response of Ag nanorod substrates. *Appl. Phys. Lett.* **2010**, *97*, 121902. [CrossRef]
33. Driskell, J.D.; Lipert, R.J.; Porter, M.D. Labeled Gold Nanoparticles Immobilized at Smooth Metallic Substrates: Systematic Investigation of Surface Plasmon Resonance and Surface-Enhanced Raman Scattering. *J. Phys. Chem. B* **2006**, *110*, 17444–17451. [CrossRef] [PubMed]
34. Mulvaney, S.P.; He, L.; Natan, M.J.; Keating, C.D. Three-layer substrates for surface-enhanced Raman scattering: Prepartion and preliminary evaluation. *J. Raman Spectrosc.* **2003**, *34*, 163–171. [CrossRef]

35. Galopin, E.; Barbillat, J.; Coffinier, Y.; Szunerits, S.; Patriarche, G.; Boukherroub, R. Silicon nanowires coated with silver nanostructures as ultrasensitive interfaces for surface-enhanced Raman spectroscopy. *ACS Appl. Mater. Interfaces* **2009**, *1*, 1396–1403. [CrossRef] [PubMed]
36. Zhang, M.L.; Fan, X.; Zhou, H.W.; Shao, M.W.; Zapien, J.A.; Wong, N.B.; Lee, S.T. A high-efficiency surface-enhanced Raman scattering substrate based on silicon nanowires array decorated with silver nanoparticles. *J. Phys. Chem. C* **2010**, *114*, 1969–1975. [CrossRef]
37. He, Y.; Su, S.; Xu, T.T.; Zhong, Y.L.; Zapien, J.A.; Li, J.; Fan, C.H.; Lee, S.T. Silicon nanowires-based highly-efficient SERS-active platform for ultrasensitive DNA detection. *Nano Today* **2011**, *6*, 122–130. [CrossRef]
38. Schmidt, M.S.; Hübner, J.; Boisen, A. Large area fabrication of leaning silicon nanopillars for surface enhanced Raman spectroscopy. *Adv. Mater.* **2012**, *24*, OP11–OP18. [CrossRef]
39. Akin, M.S.; Yilmaz, M.; Babur, E.; Ozdemir, B.; Erdogan, H.; Tamer, U.; Demirel, G. Large area uniform deposition of silver nanoparticles through bio-inspired polydopamine coating on silicon nanowire arrays for pratical SERS applications. *J. Mater. Chem. B* **2014**, *2*, 4894–4900. [CrossRef]
40. Bryche, J.-F.; Bélier, B.; Bartenlian, B.; Barbillon, G. Low-cost SERS substrates composed of hybrid nanoskittles for a highly sensitive sensing of chemical molecules. *Sens. Actuators B* **2017**, *239*, 795–799. [CrossRef]
41. Magno, G.; Bélier, B.; Barbillon, G. Gold thickness impact on the enhancement of SERS detection in low-cost Au/Si nanosensors. *J. Mater. Sci.* **2017**, *52*, 13650–13656. [CrossRef]
42. Magno, G.; Bélier, B.; Barbillon, G. Al/Si nanopillars as very sensitive SERS substrates. *Materials* **2018**, *11*, 1534. [CrossRef] [PubMed]
43. Barbillon, G. Fabrication and SERS Performances of Metal/Si and Metal/ZnO Nanosensors: A Review. *Coatings* **2019**, *9*, 86. [CrossRef]
44. Lagarkov, A.; Boginkaya, I.; Bykov, I.; Budashov, I.; Ivanov, A.; Kurochkin, I.; Ryzhikov, I.; Rodionov, I.; Sedova, M.; Zverev, A.; et al. Light localization and SERS in tip-shaped silicon metasurface. *Opt. Express* **2017**, *25*, 17021–17038. [CrossRef] [PubMed]
45. Sarychev, A.K.; Ivanov, A.; Lagarkov, A.; Barbillon, G. Light Concentration by Metal-Dielectric Micro-Resonators for SERS Sensing. *Materials* **2019**, *12*, 103. [CrossRef]
46. Kanipe, K.N.; Chidester, P.P.F.; Stucky, G.D.; Meinhart, C.D.; Moskovits, M. Properly Structured, Any Metal Can Produce Intense Surface Enhanced Raman Spectra. *J. Phys. Chem. C* **2017**, *121*, 14269–14273. [CrossRef]
47. Whelan, C.M.; Smyth, M.R.; Barnes, C.J. HREELS, XPS, and Electrochemical Study of Benzenethiol Adsorption on Au(111). *Langmuir* **1999**, *15*, 116–126. [CrossRef]
48. Lu, G.; Xu, J.; Wen, T.; Zhang, W.; Zhao, J.; Hu, A.; Barbillon, G.; Gong, Q. Hybrid Metal-Dielectric Nano-Aperture Antenna for Surface Enhanced Fluorescence. *Materials* **2018**, *11*, 1435. [CrossRef]
49. Zambrana-Puyalto, X.; Ponzellini, P.; Maccaferri, N.; Tessarolo, E.; Pelizzo, M.G.; Zhang, W.; Barbillon, G.; Lu, G.; Garoli, D. A hybrid metal-dielectric zero mode waveguide for enhanced single molecule detection. *Chem. Commun.* **2019**, *55*, 9725–9728. [CrossRef]
50. Colas, F.; Barchiesi, D.; Kessentini, S.; Toury, T.; Lamy de la Chapelle, M. Comparison of adhesion layers of gold on silicate glasses for SERS detection. *J. Opt.* **2015**, *17*, 114010. [CrossRef]
51. Johnson, P.B.; Christy, R.W. Optical Constants of the Noble Metals. *Phys. Rev. B* **1972**, *6*, 4370–4379. [CrossRef]
52. McPeak, K.M.; Jayanti, S.V.; Kress, S.J.P.; Meyer, S.; Iotti, S.; Rossinelli, A.; Norris, D.J. Plasmonic Films Can Easily Be Better: Rules and Recipes. *ACS Photonics* **2015**, *2*, 326–333. [CrossRef] [PubMed]
53. Schinke, C.; Peest, P.C.; Schmidt, J.; Brendel, R.; Bothe, K.; Vogt, M.R.; Kröger, I.; Winter, S.; Schirmacher, A.; Lim, S.; et al. Uncertainty analysis for the coefficient of band-to-band absorption of crystalline silicon. *AIP Adv.* **2015**, *5*, 067168. [CrossRef]
54. Manjare, M.; Wang, F.; Rodrigo, S.G.; Harutyunyan, H. Exposing optical near fields of plasmonic patch nanoantennas. *Appl. Phys. Lett.* **2017**, *111*, 221106. [CrossRef]
55. Scalora, M.; Vincenti, M.A.; de Ceglia, D.; Grande, M.; Haus, J.W. Raman scattering near metal nanostructures. *J. Opt. Soc. Am. B* **2012**, *29*, 2035–2045. [CrossRef]
56. Tetsassi Feugmo, C.G.; Liégeois, V. Analyzing the vibrational signatures of thiophenol adsorbed on small gold clusters by DFT calculations. *ChemPhysChem* **2013**, *14*, 1633–1645. [CrossRef]
57. Li, S.; Wu, D.; Xu, X.; Gu, R. Theoretical and experimental studies on the adsorption behavior of thiophenol on gold nanoparticles. *J. Raman Spectrosc.* **2007**, *38*, 1436–1443. [CrossRef]

58. Temple, P.A.; Hathaway, C.E. Multiphonon Raman Spectrum of Silicon. *Phys. Rev. B* **1973**, *7*, 3685–3697. [CrossRef]
59. Khorasaninejad, M.; Walia, J.; Saini, S.S. Enhanced first-order Raman scattering from arrays of vertical silicon nanowires. *Nanotechnology* **2012**, *23*, 275706. [CrossRef]
60. Stern, D.A.; Wellner, E.; Salaita, G.N.; Laguren-Davidson, L.; Lu, F.; Batina, N.; Frank, D.G.; Zapien, D.C.; Walton, N.; Hubbard, A.T. Adsorbed Thiophenol and Related-Compounds Studied at Pt(111) Electrodes by EELS, Auger Spectroscopy and Cyclic Voltammetry. *J. Am. Chem. Soc.* **1988**, *110*, 4885–4893. [CrossRef]
61. Caldwell, J.D.; Glembocki, O.; Bezares, F.J.; Bassim, N.D.; Rendell, R.W.; Feygelson, M.; Ukaegbu, M.; Kasica, R.; Shirey, L.; Hosten, C. Plasmonic Nanopillar Arrays for Large-Area, High-Enhancement Surface-Enhanced Raman Scattering Sensors. *ACS Nano* **2011**, *5*, 4046–4055. [CrossRef]
62. Alvarez-Puebla, R.A. Effects of the Excitation Wavelength on the SERS Spectrum. *J. Phys. Chem. Lett.* **2012**, *3*, 857–866. [CrossRef] [PubMed]
63. Le Ru, E.C.; Blackie, E.J.; Meyer, M.; Etchegoin, P.G. Surface enhanced Raman scattering enhancement factors: A comprehensive study. *J. Phys. Chem. C* **2007**, *111*, 13794–13803. [CrossRef]
64. Zheng, M.; Zhu, X.; Chen, Y.; Xiang, Q.; Duan, H. Three-dimensional donut-like gold nanorings with multiple hot spots for surface-enhanced Raman spectroscopy. *Nanotechnology* **2017**, *28*, 0453503. [CrossRef] [PubMed]
65. Etchegoin, P.G.; Le Ru, E.C. Basic Electromagnetic Theory of SERS. In *Surface Enhanced Raman Spectroscopy: Analytical, Biophysical and Life Science Applications*; Schlücker, S., Ed.; Wiley-VCH: Weinheim, Germany, 2011; pp. 1–34, ISBN 978-3-527-32567-2.

© 2019 by the authors. Licensee MDPI, Basel, Switzerland. This article is an open access article distributed under the terms and conditions of the Creative Commons Attribution (CC BY) license (http://creativecommons.org/licenses/by/4.0/).

Review

Latest Novelties on Plasmonic and Non-Plasmonic Nanomaterials for SERS Sensing

Grégory Barbillon

EPF-Ecole d'Ingénieurs, 3 bis rue Lakanal, 92330 Sceaux, France; gregory.barbillon@epf.fr

Received: 4 June 2020; Accepted:16 June 2020; Published: 19 June 2020

Abstract: An explosion in the production of substrates for surface enhanced Raman scattering (SERS) has occurred using novel designs of plasmonic nanostructures (e.g., nanoparticle self-assembly), new plasmonic materials such as bimetallic nanomaterials (e.g., Au/Ag) and hybrid nanomaterials (e.g., metal/semiconductor), and new non-plasmonic nanomaterials. The novel plasmonic nanomaterials can enable a better charge transfer or a better confinement of the electric field inducing a SERS enhancement by adjusting, for instance, the size, shape, spatial organization, nanoparticle self-assembly, and nature of nanomaterials. The new non-plasmonic nanomaterials can favor a better charge transfer caused by atom defects, thus inducing a SERS enhancement. In last two years (2019–2020), great insights in the fields of design of plasmonic nanosystems based on the nanoparticle self-assembly and new plasmonic and non-plasmonic nanomaterials were realized. This mini-review is focused on the nanoparticle self-assembly, bimetallic nanoparticles, nanomaterials based on metal-zinc oxide, and other nanomaterials based on metal oxides and metal oxide-metal for SERS sensing.

Keywords: SERS; sensors; plasmonics; zinc oxide; metal oxides; self-assembly; bimetallic nanoparticles

---

1. Introduction

The strong development of plasmonic nanomaterials for various applications such as photovoltaics [1–4], optical devices [5–10], and biochemical sensors [11–17] has taken place over these last ten years. The plasmonic nanostructures can also enable the detection of phase transitions under high-pressure conditions [18], the luminescence upconversion enhancement [19,20], and the optical tuning of photoluminescence [21] and upconversion luminescence [22]. For plasmonic sensors of biomolecules, the surface enhanced Raman scattering (SERS) is largely employed as a very sensitive technique of analysis. For maximizing the enhancement factor (EF) of SERS signal, the electromagnetic contribution is predominantly used. EF is calculated by taking the fourth power of the electric field amplitude obtained with the plasmonic nanostructures [23]. The key point in order to obtain zones of strong electric field (called hotspots) is a precise control of the shape, size, and spatial organization of plasmonic nanostructures. The control of these parameters is enabled and realized thanks to a great number of lithographies such electron beam lithography [24–27], optical lithographies [28–30], nanosphere lithography [31–33], and nanoimprint lithography [34–36]. Several groups examined a broad number of designs as plasmonic nanodisks, nanodimers, and nanorods, which have reached important EF values (EF = 10^6–10^9) [37–39]. Furthermore, a gain of 1 or 2 orders of magnitude on the enhancement factor can be realized by inserting a metallic layer under the plasmonic nanosystems (EF = 10^6–10^9). A coupling between the nanosystems by means of surface plasmon polaritons on the metallic film [40,41] or hybridization of localized plasmon modes with the image modes in a plasmonic substrate [42,43] allows this gain. Moreover, this advanced type of SERS substrate can allow an application to multispectral SERS sensing [44]. Another approach for obtaining an excellent SERS activity is to use luminescent-plasmonic material based on neodymium(III)-doped yttrium–aluminium–silicate microspheres with gold nanoparticles [45]. In addition, hybrid metal/Si nanostructures allowed achieving substantial values of EF (10^7–10^{10}) [46–52]. These hybrid nanostructures based on silicon (semiconductor) have the property of biocompatibility, and a low cost of production. Moreover, they permit the emergence of hotspots placed at the level of the interface of the metal and semiconductor. Furthermore, another possible outcome is based on the zinc oxide (ZnO) nanostructures capped with metallic layer or metallic nanoparticles in order to achieve excellent enhancement factors (EF = 10^6–10^{10}) [53–56]. The use of bimetallic nanosystems offers the possibility to have excellent functionalities concerning the plasmonic and chemical properties compared to plasmonic nanosystems composed of an unique metal [57,58]. As silver has a better plasmonic enhancement than gold, the bimetallic gold-silver nanosystems are developed in order to suppress oxidation of silver with gold [59]. Thus, sharper and stronger characteristics of localized surface plasmon resonances (LSPRs) for the bimetallic systems enable obtaining larger SERS activities due to the hotspots coming from the LSPR coupling between Au and Ag nanosystems [60,61]. Another way is to design effective SERS substrates by self-assembling of plasmonic nanoparticles [62,63]. The advantages of the self-assembly are the low cost and time of fabrication of SERS substrates. This effectiveness is strongly depending on the distance between plasmonic nanoparticles [64]. Nevertheless, the reproducibility of the SERS signal is very weak with this type of substrates [65]. However, improvements have emerged as the template-assisted self-assembly suppressing the issue of signal reproducibility [66,67]. In addition, alternative materials such as metal oxides (different of zinc oxide as MoO_3 molybdenum trioxide, WO_{3-x} tungsten oxide or $CoFe_2O_4$ cobalt ferrite) emerged for SERS application [68,69].

The aim of this mini-review is to present the latest novelties on plasmonic and non-plasmonic nanomaterials for SERS sensing over the period 2019–2020. We will focus on the self-assembly of plasmonic nanoparticles in a first part. Then, bimetallic nanosystems will be addressed, then nanomaterials based on metal-ZnO, and finally other nanomaterials based on metal oxides and metal oxide-metal.

2. Novelties on Plasmonic and Non-Plasmonic Nanomaterials for SERS Sensing

In order to compare the different SERS performances for all the nanosystems presented in this mini-review, the detection limits (LODs) obtained experimentally were used, and also the calculation of the enhancement factor (EF) [37] or the analytical enhancement factor (AEF) [39] (see tables of each section). The formulas of EF and AEF were expressed as follows:

$$EF = \frac{I_{SERS}}{I_{Raman}} \times \frac{N_{Raman}}{N_{SERS}} \qquad (1)$$

$$AEF = \frac{I_{SERS}}{I_{Raman}} \times \frac{C_{Raman}}{C_{SERS}} \qquad (2)$$

where I_{SERS}, I_{Raman} represent the SERS and Raman intensities, respectively. N_{SERS}, N_{Raman}, C_{SERS}, C_{Raman} are the numbers and concentrations of analyte molecules for SERS and reference Raman experiments, respectively.

2.1. SERS Substrates Designed by Self-Assembly

Novel SERS substrates were designed by self-assembly in the period 2019–2020 (see Table 1). The first example concerns the fabrication of nanogap plasmonic micropillars by using the capillary-force driven self-assembly (CFSA). These SERS substrates allowed achieving enhancement factors up to 8×10^7 in a fluidic medium. Moreover, a detection limit (LOD) of 0.1 mM for doxorubicin (DOX = anticancer drug) was reached with this type of structures. This fabrication method was very flexible, because it allowed realizing plasmonic structures on flat and non-flat substrates [70]. Ghosh et al. showed that plasmonic dimers with subnanometer gap enabled to reach enhancement factor of 10^7 and a Rhodamine 6G (R6G) detection at the ppb level. These plasmonic nanostructures were realized by directed microwave-assisted self-assembly and segregated by a graphene monolayer [71].

Table 1. Surface enhanced Raman scattering (SERS) performances of substrates designed by self-assembly for biological/chemical sensing (LOD = limit of detection; DOX = doxorubicin; 4-ATP = 4-aminothiophenol; PVC = polyvinyl chloride; 4-NTP = 4-nitrothiophenol).

SERS Substrates	Detected Molecules	EF or AEF	LOD	References
3D Au Nanogap micropillars	Rhodamine 6G	8×10^7	–	[70]
3D Au Nanogap micropillars	DOX	–	0.1 mM	[70]
Au-Graphene-Au dimers	Rhodamine 6G	10^7	ppb level	[71]
Au nanorods with Au spheres	4-NTP	10^4–10^5	–	[72]
Au nanoislands with disorder control	Rhodamine 6G	10^7–10^8	1 nM	[73]
AuNPs/PVC film	4-ATP	3.7×10^6	–	[74]
AuNPs/PVC film	thiram	–	10 ng·cm^{-2}	[74]
Bimetallic array with Au microrings	4-ATP	4.2×10^5	1 nM	[75]

Kuttner et al. reported on SERS performances obtained with gold nanorods on which were deposited self-assembled Au nanospheres (see Figure 1). High AEF values of 10^4–10^5 were achieved due to the coupling of plasmonic modes of gold nanorods and gold nanospheres (see Figure 1). The 4-nitrothiophenol (4-NTP) molecules were used in order to determine the AEF value [72]. From Figure 1, it has been also observed that the AEF values were higher of a magnitude order for the longitudinal (C_L) coupled mode than the transversal (C_T) coupled mode. This was due to the fact that the excitation wavelength had a better position in comparison to the resonance position of the C_L mode [72].

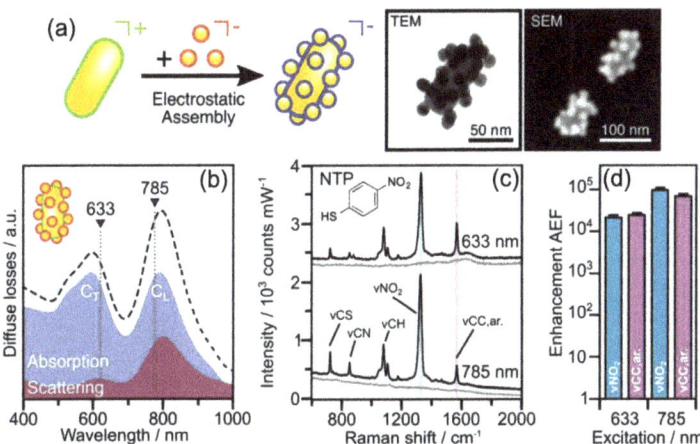

Figure 1. (a) Scheme of the self-assembly of Au nanospheres on Au nanorods. TEM and SEM images of the obtained superstructures. (b) Absorption and scattering spectra of the plasmonic superstructures with the chosen wavelengths of excitation for the transversal (C_T) and longitudinal (C_L) coupled modes, which are 633 nm and 785 nm, respectively. (c) SERS spectra of plasmonic superstructures without (in grey) and with 4-NTP molecules (C_{NTP} = 1 µM, in black) for the two wavelengths of excitation. (d) Analytical enhancement factor (AEF) values corresponding to the SERS spectra in (c). All the figures are reproduced from [72] with permission from the Royal Society of Chemistry.

Fusco et al. discussed SERS performances of gold nano-island (NI) substrates obtained by self-assembly of which the disorder degree of NIs was controllable. The reached EF values were in the range of 10^7–10^8 and the lowest concentration detected was 1 nM with the R6G molecules [73]. Wu et al. showed an one-step method of fabrication of SERS substrates constituted of gold nanoparticles (AuNPs) and a polyvinyl chloride (PVC) film via an interfacial self-assembly induced by polymer. This fabrication of these AuNPs/PVC films was simple, low cost and these films can be reused. Moreover, the SERS performances were excellent such as an EF of 3.7×10^6 for the sensing of 4-aminothiophenol (4-ATP) molecules, and a LOD of 10 ng·cm^{-2} for pesticides (thiram) [74]. To conclude this section, Yin et al. fabricated bimetallic arrays composed of gold micro-rings, which decorated platinum (Pt) disks (see Figure 2) by employing a process of templated-self-assembly (TSA). Thus, a SERS enhancement of 4.2×10^5 was attained for the sensing of 4-aminothiophenol molecules (4-ATP). The limit of detection of 4-ATP obtained with these superstructures was equal to the concentration of 1 nM (see Figure 2) [75].

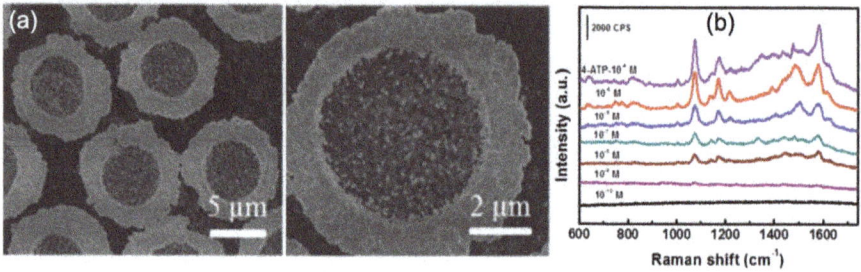

Figure 2. (a) SEM images of the obtained Pt disks decorated by gold micro-rings. (b) SERS spectra of 4-ATP molecules for different concentrations. All the figures are reproduced from [75] with permission from the Royal Society of Chemistry.

2.2. Bimetallic Nanoparticles for SERS Sensing

Recent advances on bimetallic nanoparticles for SERS sensing have occurred over the period of 2019–2020 (see Table 2). Firstly, Su et al. reached an improved sensitivity of detection of cardiorenal syndrome markers by using the combination of 3D ordered macroporous Au-Ag-Au array (substrate) and Ag-Au nanostars (nanotags). Thus, this combination enabled generating hotspots coming from the plasmonic coupling in near-field inducing a SERS enhancement. The LODs observed experimentally with this plasmonic nanosystem (substrate+nanotags) were 0.41, 0.53, and 0.76 fg·mL^{-1} for neutrophil gelatinase-associated lipocalin (NGAL), N-terminal prohormone of brain natriuretic peptide (NT-ProBNP), and cardiac troponin (cTnI), respectively [76]. Ning et al. also demonstrated SERS performances obtained by using Au–Ag–Ag nanorod coupled to the magnetic beads through DNA hybridization. A LOD of 1 fM for HPV-16 fragments (human papillomavirus DNA type 16) was found [77]. Hussain et al. reported on the quick contaminant detection in milk by using Au/Ag core-shell nanoparticles. LODs of 0.21 and 14.88 ppm for thiram and dicyandiamide (DCD) were obtained in milk, respectively. The detection limits of molecules of interest with these plasmonic Au/Ag nanoparticles can be thus determined in a short time (34 min) by using the fabrication approach proposed in this paper [78]. Tian et al. showed the easy synthesis of Au/Ag nanoparticles rich in silver by using the combination of galvanic replacement process and co-reduction of silver atoms. These bimetallic nanoparticles enabled obtaining an improved SERS activity caused by the great presence of silver in the nanoparticles. In order to evaluate the SERS activity of these Au/Ag nanoparticles rich in silver, the authors chose using thiophenol molecules, and an EF value of 2.3×10^6 was found. Then, the authors labeled their Au/Ag nanoparticles with Atto-610 antibodies and added gold nanoparticles through electrostatic adsorption for the SERS detection of rabbit IgG. Thus, the authors achieved a LOD of 20 pg·L^{-1} for rabbit IgG [79].

Table 2. SERS performances of bimetallic nanoparticles for biological/chemical sensing (4-MBA = 4-mercaptobenzoic acid; cTnI = cardiac troponin; NT-ProBNP = N-terminal prohormone of brain natriuretic peptide; NGAL = neutrophil gelatinase-associated lipocalin; HPV-16 = human papillomavirus DNA type 16; DCD = dicyandiamide; 4-MPY = Mercaptopyridine; MB = Methylene blue; 4-MPh = 4-mercaptothiophenol; PDOP = Polydopamine).

SERS Substrates	Detected Molecules	EF or AEF	LOD	References
3DOM Au-Ag-Au array with Ag-Au stars	cTnI	–	0.76 fg·mL^{-1}	[76]
3DOM Au-Ag-Au array with Ag-Au stars	NT-ProBNP	–	0.53 fg·mL^{-1}	[76]
3DOM Au-Ag-Au array with Ag-Au stars	NGAL	–	0.41 fg·mL^{-1}	[76]
Au@AgAg nanorods	HPV-16	–	1 fM	[77]
Au@AgNPs	thiram	–	0.21 ppm	[78]
Au@AgNPs	DCD	–	14.88 ppm	[78]
Au@AgNPs	rabbit IgG	–	20 pg·L^{-1}	[79]
Au@AgNPs	Thiophenol	2.3×10^6	–	[79]
Potato shaped Au-Ag NPs	MB	–	1 fM	[80]
AuAg@Ag hollow cubic NSs	4-MPh	–	1 aM	[81]
Ag-Au@NF	Rhodamine 6G	–	0.1 nM	[82]
AuNP@PDOP@AgNP	MB	3.5×10^5	–	[83]
Xylan-capped Au@Ag	Sudan I	–	1 nM	[84]
Xylan-capped Au@Ag	4-MBA	1.24×10^8	1 nM	[84]
Au@AgNPs	4-MPY	3.5×10^7	<1 nM	[85]

William et al. demonstrated SERS performances of sprouted potato-shaped bimetallic nanoparticles. The shape of these Au/Ag nanoparticles was obtained by carefully setting the quantity of silver for a given quantity of gold. Thus, the authors obtained a LOD of 1 fM for methylene blue molecules [80]. Joseph et al. reported on the fabrication AuAg@Ag hollow cubic nanosystems for a detection of mercaptothiophenol (4-MPh) molecules. These structures were composed of an AuAg core and an Ag shell. A LOD of 1 aM was found for 4-MPh molecules. This efficiency of these SERS nanostructures was caused by the electromagnetic and chemical contributions. The major part of this efficiency was due to the charge transfer of 4-MPh molecules via the silver shell to the alloy core [81]. Vu et al. realized Ag-Au nanostructures on nickel foam as SERS substrates. Authors employed rhodamine 6G molecules in order to test their 3D nanostructures, and a LOD of 0.1 nM was found for these molecules. Moreover, the SERS signal was durable even after 100 cycles of abrasion with sandpaper, or after sonication for half an hour for these Ag-Au nanostructures on nickel foam [82]. Yilmaz et al. demonstrated a SERS activity with bimetallic core–shell nanoparticles with an intermediate layer of bioinspired polydopamine between Au and Ag nanoparticles. An EF value of 3.5×10^5 for the detection of methylene blue molecules was found. This bioinspired polydopamine layer is employed as stabilizing agent for adsorption of silver nanoparticles as well as reducing agent for reduction of Ag ions [83].

Furthermore, Cai et al. demonstrated the green synthesis Au–Ag core-shell nanoparticles by using xylan for SERS sensing (see Figure 3). With these xylan-capped Au@Ag nanoparticles, a detection limit of 1 nM and an AEF value of 1.24×10^8 for 4-mercaptobenzoic acid (4-MBA) molecules were found as well as a LOD of 1 nM for Sudan I molecules (food contaminant; see Figure 3). These SERS performances were explained by the fact that the xylan capping allowed creating hotspots between Au/Ag nanoparticles. Optimal SERS performances for xylan-capped Au/Ag nanoparticles were realized for a mole ratio of AgNO$_3$ to HAuCl$_4$ equal to 4 and a dosage of xylan equal to 1 in the fabrication of Ag shell corresponding to 2.86×10^{-8} mol of xylan (see Figure 3) [84].

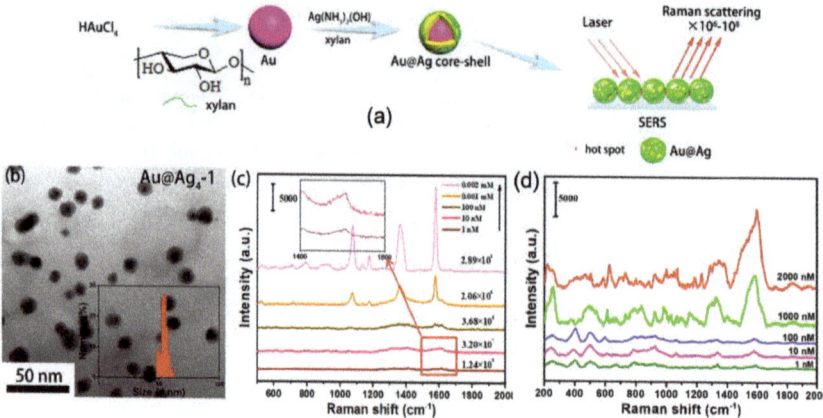

Figure 3. (a) Principle scheme of the synthesis of Au/Ag core-shell nanoparticles by using xylan for SERS sensing. (b) TEM image of xylan-capped Au@Ag nanoparticles with optimized parameters for SERS. SERS spectra of (c) 4-MBA molecules and (d) Sudan I molecules for different concentrations obtained with optimized xylan-capped Au/Ag nanoparticles of (b). All the figures are reprinted (adapted) with permission from [84], Copyright 2019 American Chemical Society.

To conclude this section on bimetallic nanosystems, Prakash et al. reported on the SERS detection of bacteria by using Au/Ag plasmonic nanoparticles which were positively charged (see Figure 4). Firstly, authors tested the SERS properties of these plasmonic nanoparticles by employing mercaptopyridine (4-MPY) molecules. An AEF of 3.5×10^7 and a LOD inferior to nanomolar concentration were found (see Figure 4). Then, the detection of bacteria (e.g., *Escherichia coli*) was realized as proof-of-concept (see Figure 4). Thus, these bimetallic nanoparticles charged positively allowed basic experimental conditions without using specific processes for SERS sensing of bacteria [85].

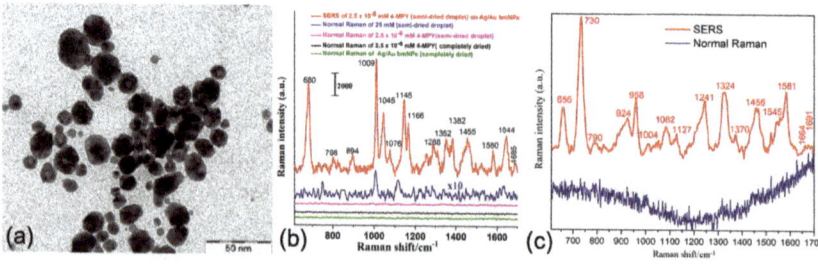

Figure 4. (a) TEM image of Au/Ag nanoparticles. (b) SERS spectrum of 4-MPY molecules at 2.5 nM concentration (in red). (c) SERS spectrum of *Escherichia coli*. All the figures are reprinted (adapted) with permission from [85], Copyright 2020 American Chemical Society.

2.3. Nanomaterials Based on Metal-ZnO for SERS Sensing

The use of zinc oxide associated to a noble metal for the fabrication of highly efficient SERS substrates increased since these last years. In this section, we present latest works on this subject over the period 2019–2020 (see Table 3). Fularz et al. reported on SERS performances of ZnO nanowires coated with Ag nanoparticles. Their idea was to treat these hybrid nanostructures by heat, which enabled an efficient charge transfer in order to enhance the SERS signal. This heat processing in oxygen environment introduced defects as interstitial oxygen in ZnO structure. This interstitial oxygen reduced optical gap favoring the charge transfer between hybrid nanowires and molecules. Moreover, the heat processing changed the wettability of Ag/ZnO nanowires (see Figure 5a,b) that had for effect of decreasing the spreading of Ag nanoparticles or studied molecules on ZnO nanowire surface which also favored the SERS enhancement. Figure 5c displays the effect of the annealing temperature of Ag/ZnO nanowires on the SERS spectra of (meso-tetra(N-methyl-4-pyridyl)porphine tetrachloride (TMPyP) molecules. A LOD of 100 nM is found for TMPyP molecules with Ag/ZnO nanowires annealed at 200 °C (see Figure 5d) [86].

Table 3. SERS performances of metal-ZnO-based nanostructures for biological/chemical sensing (AgNPs = Ag nanoparticles; TMPyP = (meso-tetra(N-methyl-4-pyridyl)porphine tetrachloride; WGM = whispering gallery mode).

SERS Substrates	Detected Molecules	EF or AEF	LOD	References
ZnO Nanowires with AgNPs	TMPyP	–	100 nM	[86]
ZnO heterostructure with AgNPs	Malachite green	–	0.1 pM	[87]
Ag/ZnO heterostructure	Methylene orange	1.3×10^{10}	1 pM	[88]
ZnO/graphene/Ag WGM microcavity	Rhodamine 6G	9.5×10^{11}	1 fM	[89]
Hollow ZnO@Ag nanospheres	Nitrite	–	3 nM	[90]
Ag/ZnO nanorods	Pioglitazone	–	1 nM	[91]
Ag/ZnO nanorods	Phenformin	–	5 nM	[91]
Ag/ZnO/Au nanorods	λ-DNA	–	0.3 nM	[92]
Au/ZnO nanorods	Dopamine	1.2×10^4	–	[93]
ZnO nanorods with AuNPs	Methylene blue	–	1 nM	[94]
Au/ZnO hollow urchins	Adenine	–	1 µM	[95]
Au/ZnO hollow urchins	Thiophenol	–	10 nM	[95]

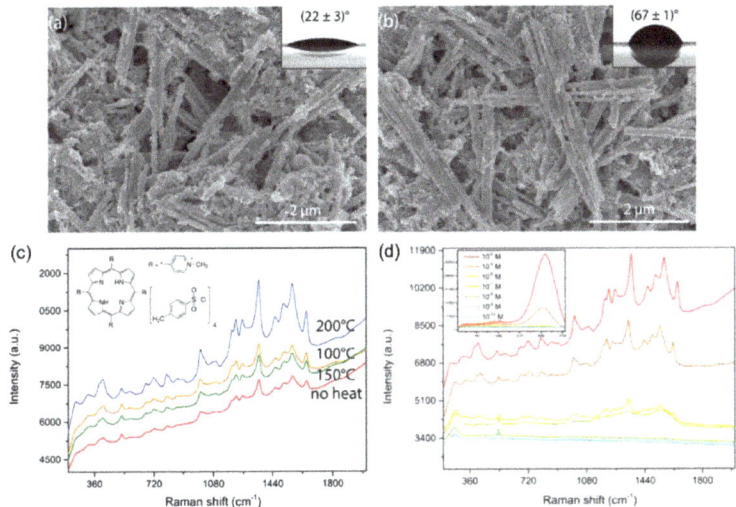

Figure 5. SEM pictures of (**a**) ZnO nanowires not annealed, and (**b**) annealed at 200 °C. Insets display contact angles for each surface. (**c**) SERS spectra of TMPyP recorded on Ag/ZnO nanowires for several temperatures. (**d**) SERS spectra of TMPyP recorded on Ag/ZnO nanowires annealed at 200 °C for various concentrations of TMPyP. Inset in (**d**) displays the whole spectrum with fluorescence signal that appears when the concentration is superior or equal to 10^{-5} M. All the figures are reprinted (adapted) with permission from [86], Copyright 2020 American Chemical Society.

The following examples present two studies reporting on the SERS performances of Ag/ZnO heterostructures. Firstly, Yao et al. investigated SERS performances of Mg-doped ZnO heterostructures coated with Ag nanoparticles. Authors demonstrated that the enhancement of SERS signal was due to the combination of electromagnetic contribution and charge transfer, and they reported a LOD of 0.1 pM for detection of malachite green molecules [87]. Secondly, Rajkumar and Sarma demonstrated excellent SERS performances obtained with Ag/ZnO heterostructures composed of ZnO microrods decorated by Ag nanoparticles. This design allowed obtaining a clustering of Ag nanoparticles inducing the formation hotspots. These hotspots enabled the enhancement of the SERS signal. Thus, LOD of 1 pM and AEF of 1.3×10^{10} were found with these heterostructures for the detection of methylene orange molecules [88]. Next, Zhu et al. reported on the use of a hybrid microcavity composed of ZnO, graphene and silver for enhancing the SERS signal. This enhancement is obtained by combining a whispering-gallery mode of a microcavity, the plasmonic resonance of Ag nanoparticles and a charge transfer between studied molecules and graphene. Thus, an EF value of 9.5×10^{11} and a LOD of 1 fM were obtained with this structure for detection of rhodamine 6G molecules [89]. Wang et al. investigated SERS performances of hollow ZnO@Ag nanospheres for detection of nitrite species. Authors demonstrated a LOD of 3 nM for detection of nitrite species [90]. In the next examples, four studies based on metal/ZnO nanorods are presented. The first one concerns Ag/ZnO nanorods. In this first investigation, authors demonstrated SERS performances due to charge transfers. LODs of 1 nM and 5 nM were found for detection of pioglitazone and phenformin, respectively [91]. Then, Pal et al. demonstrated that a bimetallic/ZnO structure composed of silver, zinc oxide, and gold allowed significant SERS performances. Authors reported an excellent LOD of 0.3 nM for detection of lambda DNA [92]. The last two examples are dedicated to the SERS performances of Au/ZnO nanorods. At first, Zhou et al. showed that Au/ZnO heterogeneous nanorods allowed an enhancement of SERS signal due to charge transfer enhanced by the localized surface plasmon resonance of a gold nanoparticle located at an extremity of ZnO nanorods. A value of the enhancement factor of 1.2×10^4 was determined for detection of dopamine [93]. Next, Doan et al. reported on the use ZnO nanorods coated with gold nanoparticles for enhancing SERS performances. Authors also demonstrated that their SERS substrates were self-cleaning under UV light, and found a LOD of 1 nM for detection of methylene blue molecules [94]. To finish this section on metal-ZnO-based nanomaterials, Graniel et al. reported on the fabrication of Au/ZnO hollow nano-urchins (see Figure 6a) and their SERS performances. These hybrid nano-urchins enabled the formation of hotspots (strong electric field zones) which induced enhancements of SERS signal. Thus, authors found LODs of 10 nM and 1 µM for detection of thiophenol and adenine, respectively (see Figure 6b,c). Moreover, they demonstrated an excellent substrate-to-substrate reproducibility with a relative standard deviation < 10% [95].

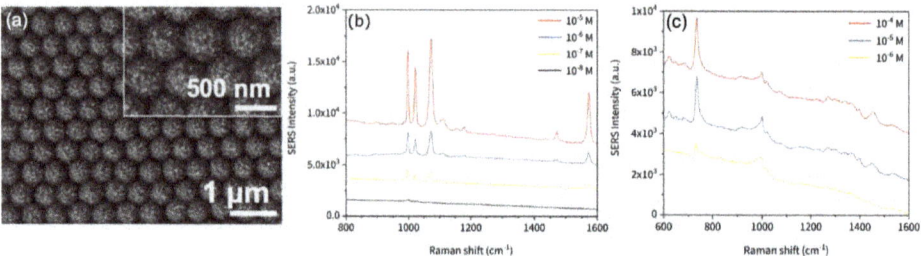

Figure 6. (a) SEM picture of Au/ZnO nano-urchins. SERS spectra of thiophenol (b) and adenine (c) molecules recorded on Au/ZnO nano-urchins for several concentrations of the studied molecules. All the figures are reproduced from [95] with permission from the Royal Society of Chemistry.

2.4. Nanomaterials Based on Metal Oxides and Based on Metal Oxide-Metal for SERS Sensing

In the previous section, we discussed the nanomaterials based on zinc oxide. In this section, recent novelties concerning to other nanomaterials based on metal oxides, then based on metal oxide-metal are presented for SERS sensing on the period of 2019–2020 (see Table 4).

Table 4. SERS performances of metal-oxide nanostructures for biological/chemical sensing (vdW MoO_3 = van der Waals molybdenum trioxide; TMOs = transition metal oxides; $W_{18}O_{49}$ = $WO_{2.72}$ = non-stoichiometric tungsten oxide; WO_{3-x} = non-stoichiometric tungsten oxide; QD = quantum dot; NW = nanowire; ZrO_2 = zirconia; $CoFe_2O_4$ = cobalt ferrite).

SERS Substrates	Detected Molecules	EF or AEF	LOD	References
Few-layered vdW MoO_3 nanosheets	Rhodamine 6G	–	20 nM	[96]
TMOs planar SERS chips	Rhodamine 6G	–	1 nM	[97]
$W_{18}O_{49}$-H_2 nanowire film	Rhodamine B	4.4×10^5	0.1 µM	[98]
WO_{3-x}-based SERS substrate	Rhodamine B	1.2×10^6	0.1 µM	[99]
WO_{3-x}QD@AgNW	Methylene blue	–	1 µM	[100]
ZrO_2@$CoFe_2O_4$@Au nanoparticles	Thiolated malachite green	5×10^{10}	–	[101]

In the first four examples, nanomaterials based on metal oxides are presented for SERS sensing. He et al. reported on the SERS performances of few-layered MoO_3 nanosheets deposited on a SiO_2/Si substrate. Authors demonstrated that SERS enhancement was due to a chemical mechanism when the thickness of these nanosheets was reduced. Moreover, this chemical mechanism was further enhanced via an atomic intercalation in the van der Walls gap. Thus, a LOD of 20 nM for detection of rhodamine 6G molecules was obtained with these MoO_3 nanosheets [96]. Hou et al. proposed an alternative strategy to classical plasmonic nanostructures for the fabrication of efficient SERS substrates. This strategy was to use non-stoichiometric transition metal oxides (TMOs) as SERS substrates. These planar TMOs SERS substrates were realized through a magnetron sputtering coupled to H_2 annealing. In this study, authors chose to investigate the following non-stoichiometric groups of TMOs: IVB, VB and VIB. They obtained a lowest LOD of 1 nM for detection of rhodamine 6G molecules. The SERS enhancement was due to the mechanism of photoinduced charge transfer from oxygen vacancies [97]. Wang et al. demonstrated the SERS performances of a $W_{18}O_{49}$-H_2 nanowire film. A LOD of 0.1 µM and an EF value of 4.4×10^5 were found for detection of rhodamine B (RhB) molecules. These SERS performances were due to the presence of oxygen vacancies in $W_{18}O_{49}$-H_2 film of which the vacancy number was increased by the reduction of H_2. This improved number of oxygen vacancies enabled enriching surface states of substrate allowing the adsorption of an increased number of RhB on this same substrate and thus inducing the enhancement of SERS activity through a charge transfer mechanism [98]. Zhou et al. demonstrated an electrical control of SERS enhancement based on tungsten oxide surface (WO_{3-x}) deposited on a SiO_2/Si substrate. This SERS improvement was realized by an electric field that introduced defects (oxygen vacancies) in the tungsten oxide surface (see the structure scheme in Figure 7a) resulting in a better charge transfer between oxide surface and the studied molecules (here, rhodamine B = RhB). A LOD of 0.1 µM and an enhancement factor of 1.2×10^6 were found for detection of rhodamine B molecules (see Figure 7b,c) with a programmed current of leakage equal to 1 mA. From Figure 7a, this value of 1 mA for the leakage current was optimal for enhancing the SERS signal. The EF value was equal to 1.2×10^6 for this current. The SERS improvement was due to a charge transfer that was favored by good alignment of the energy levels of oxygen vacancies with the molecular energy levels of rhodamine B (see Figure 7d) [99].

In the last two examples, nanomaterials based on metal oxide-metal are presented for SERS sensing. In the first one, Wei et al. reported on the SERS performances of Ag nanowires decorated with WO_{3-x} quantum dots (WO_{3-x} QD/AgNW). The SERS activity was investigated by using methylene blue (MB) molecules (MB concentration used is 1 µM that was taken as the LOD here). Authors demonstrated that WO_3 QD/AgNW films had a better SERS activity than $WO_{2.72}$ QD/AgNW films when no irradiation with Xe lamp was applied. On the contrary, $WO_{2.72}$ QD/AgNW films had a better SERS activity than WO_3 QD/AgNW films when irradiation with Xe lamp was applied. Authors observed a decreasing of SERS activity when the content of WO_3 QDs was increased, and stated that the localized surface plasmon resonance along AgNWs was blocked by the presence of WO_3 QDs. Authors observed the same behavior with $WO_{2.72}$ QD/AgNW films. Moreover, when WO_3 QD/AgNW films were irradiated, the SERS activity was decreased due to the photo-decomposition of methylene blue molecules. However, a contrary effect was observed with $WO_{2.72}$ QD/AgNW films. Indeed, the SERS activity was improved when the irradiation time was increased. This was due to the presence of oxygen defects in $WO_{2.72}$ QDs which favored charge transfers (electrons) inducing the SERS enhancement [100].

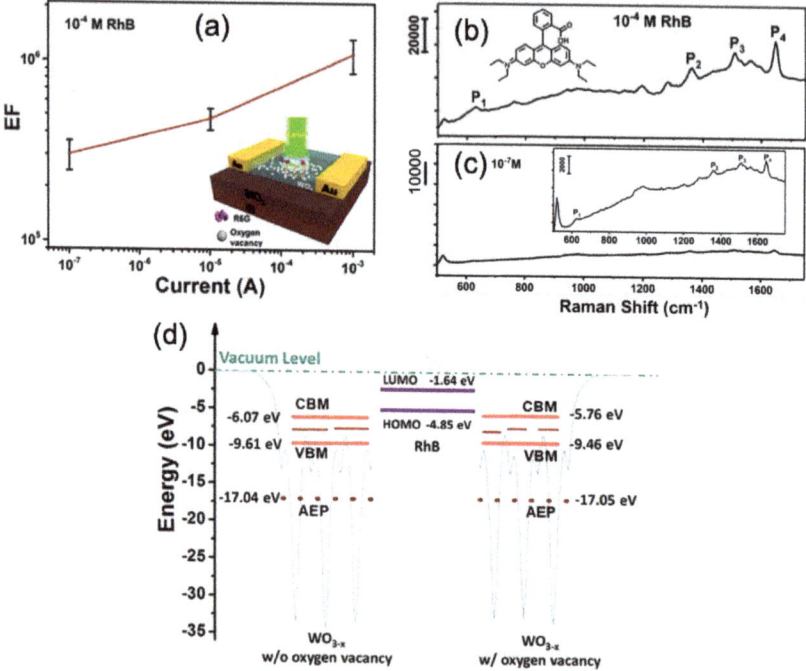

Figure 7. (**a**) Enhancement factor for SERS signal versus leakage current for a rhodamine B (RhB) concentration of 0.1 mM, and the inset represents the structure scheme. SERS spectra of RhB recorded for a leakage current of 1 mA with an RhB concentration of (**b**) 0.1 mM and (**c**) 0.1 µM (at detection limit (LOD)), and the insets of the figures (**b**) and (**c**) represent a molecular scheme of RhB and a zoom on the SERS spectrum, respectively. (**d**) Illustration and alignment of energy levels of WO_{3-x} film without oxygen vacancies (at left), RhB molecule (at center), and WO_{3-x} film with oxygen vacancies (at right). All the figures are reprinted (adapted) with permission from [99], Copyright 2019 American Chemical Society.

In the last example concluding this section, Del Tedesco et al. reported on the use of magnetoplasmonic nanoparticles for enhancing the SERS signal and its separation effect of magnetic and non-magnetic systems. In this study, a magnetoplasmonic nanoparticle was composed of a mesoporous nanoparticle of ZrO_2 on which cobalt ferrite ($CoFe_2O_4$) nanoparticles were deposited, then they were functionalized with gold nanoparticles (see Figure 8a). Other nanoparticles (NPs) were realized and only composed of a mesoporous nanoparticle of ZrO_2 on which gold nanoparticles were deposited in order to investigate the separation effect of magnetic and non-magnetic systems. Authors calculated an enhancement factor of around 5×10^{10} for SERS signal with these magnetoplasmonic nanoparticles. Moreover, the magnetoplasmonic and non-magnetic plasmonic nanoparticles were functionalized with thiolated malachite green (MG) and thiolated texas red (TR), respectively (see Figure 8b,c). Thus, authors demonstrated the separation of magnetoplasmonic and non-magnetic plasmonic nanoparticles with magnetic sorting (see Figure 8d–f). Indeed, Figure 8d displays the SERS spectrum of the solution of the mixture of the MG-functionalized magnetoplasmonic (Raman peak in green) and TR-functionalized plasmonic nanoparticles (Raman peaks in red). The SERS spectrum displayed in Figure 8e corresponds to this recorded after magnetic attraction and re-dispersed in water. The last SERS spectrum shown in Figure 8f corresponds to this recorded with the starting solution (mixture of two types of NPs) after magnetic attraction, where it only remained that the red Raman peaks corresponding to non-magnetic plasmonic nanoparticles, and also with a small remaining SERS signal (in green) corresponding to MG-functionalized magnetoplasmonic nanoparticles [101].

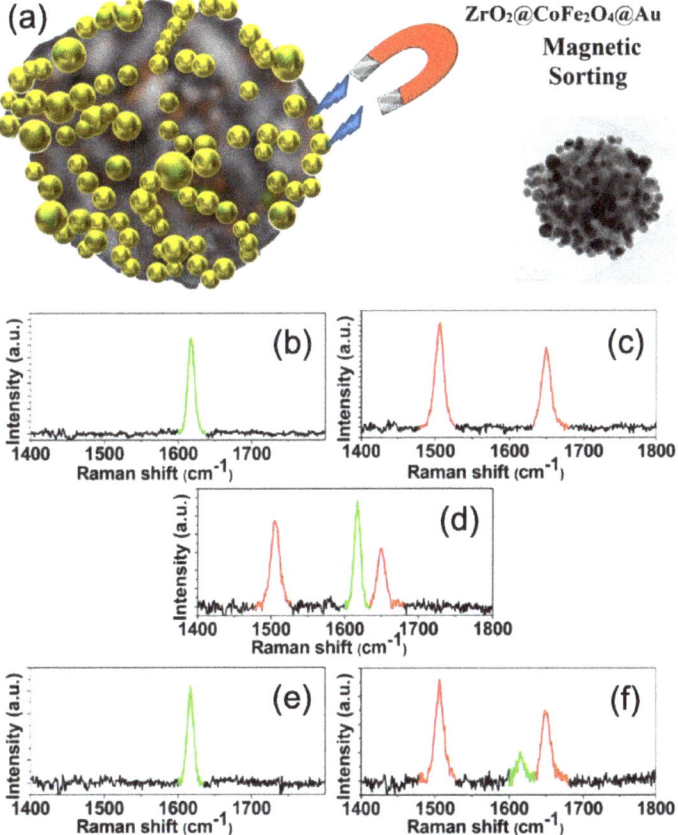

Figure 8. (a) Scheme and TEM picture of a magnetoplasmonic nanoparticle. SERS spectra of (b) malachite green on magnetoplasmonic NPs, (c) texas red on non-magnetic plasmonic NPs, (d) the mixture of (b) and (c) in solution. (e) SERS spectrum recorded after magnetic attraction from the mixed solution and re-dispersed in water. (f) SERS spectrum recorded after magnetic attraction with the mixed solution. All the figures are reprinted (adapted) with permission from [101], Copyright 2020 American Chemical Society.

3. Conclusions

In this short review, recent novelties on plasmonic and non-plasmonic nanomaterials for SERS sensing were summarized in four major parts: (i) self-assembly of plasmonic nanoparticles, (ii) bimetallic nanosystems, (iii) nanomaterials based on metal-zinc oxide, and (iv) nanomaterials based on metal oxides and metal oxide-metal. From these nanomaterials, excellent SERS performances have been obtained thanks to the generation of hotspots or an improved charge transfer. Thus, enhancement factors were in the range of 10^4–10^8, 10^5–10^8, 10^4–10^{12}, and 10^5–10^{10} for part (i), (ii), (iii) and (iv), respectively. For LOD, the values were in the range of 1 nM–0.1 mM, 1 aM–1 nM, 1 fM–1 µM, and 1 nM–0.1 µM for part (i), (ii), (iii), and (iv), respectively. By taking into account these different values of EF and LOD, the best SERS nanomaterials are bimetallic nanosystems, and nanostructures based on metal-zinc oxide, even if other nanomaterials based on metal oxides and metal oxide-metal are also good potential candidates. Moreover, this type of the fabrication strategy and nanomaterials also allowed quick, low-cost, reproducible generation of efficient SERS substrates and SERS nanotags. However, the physical/chemical properties of SERS substrates must be optimized, as coupling of the molecules with plasmonic surface, preferably in hotspots, for instance. All these properties can be optimized by using numerical simulations and experimental measurements, which are essential for acquiring a deeper understanding of all key points and achieving an efficient transfer of SERS as a regular analytical technique in the near future.

Funding: This research received no external funding.

Conflicts of Interest: The authors declare no conflict of interest.

References

1. Qin, P.; Wu, T.; Wang, Z.; Xiao, L.; Ma, L.; Ye, F.; Xiong, L.; Chen, X.; Li, H.; Yu, X.; et al. Grain Boundary and Interface Passivation with Core–Shell Au@CdS Nanospheres for High-Efficiency Perovskite Solar Cells. *Adv. Funct. Mater.* **2020**, *30*, 1908408. [CrossRef]
2. Panigrahi, S.; Jana, S.; Calmeiro, T.; Nunes, D.; Deuermeier, J.; Martins, R.; Fortunato, E. Mapping the space charge carrier dynamics in plasmon-based perovskite solar cells. *J. Mater. Chem. A* **2019**, *7*, 19811–19819. [CrossRef]
3. Cho, S.H.; Lee, J.; Lee, M.J.; Kim, H.J.; Lee, S.-M.; Choi, K.C. Plasmonically Engineered Textile Polymer Solar Cells for High-Performance, Wearable Photovoltaics. *ACS Appl. Mater. Interfaces* **2019**, *11*, 20864–20872. [CrossRef] [PubMed]
4. Yao, K.; Zhong, H.; Liu, Z.; Xiong, M.; Leng, S.; Zhang, J.; Xu, Y.-X.; Wang, W.; Zhou, L.; Huang, H.; et al. Plasmonic Metal Nanoparticles with Core–Bishell Structure for High-Performance Organic and Perovskite Solar Cells. *ACS Nano* **2019**, *13*, 5397–5409. [CrossRef] [PubMed]
5. Ono, M.; Hata, M.; Tsunekawa, M.; Nozaki, K.; Sumikura, H.; Chiba, H.; Notomi, M. Ultrafast and energy-efficient all-optical switching with graphene-loaded deep-subwavelength plasmonic waveguides. *Nat. Photonics* **2020**, *14*, 37–43. [CrossRef]
6. Zhang, X.; Liu, S.; Tan, D.; Xian, Y.; Zhang, D.; Zhang, Z.; Liu, Y.; Liu, X.; Qiu, J. Photochemically Derived Plasmonic Semiconductor Nanocrystals as an Optical Switch for Ultrafast Photonics. *Chem. Mater.* **2020**, *32*, 3180–3187. [CrossRef]
7. Guo, X.D.; Liu, R.N.; Hu, D.B.; Hu, H.; Wei, Z.; Wang, R.; Dai, Y.Y.; Cheng, Y.; Chen, K.; Liu, K.H.; et al. Efficient All-Optical Plasmonic Modulators with Atomically Thin Van Der Waals Heterostructures. *Adv. Mater.* **2020**, *32*, 1907105. [CrossRef]
8. Farmakidis, N.; Youngblood, N.; Li, X.; Tan, J.; Swett, J.L.; Cheng, Z.G.; Wright, C.D.; Pernice, W.H.P.; Bhaskaran, H. Plasmonic nanogap enhanced phase-change devices with dual electrical-optical functionality. *Sci. Adv.* **2019**, *5*, eaaw2687. [CrossRef]
9. Barbillon, G. Plasmonics and its Applications. *Materials* **2019**, *12*, 1502. [CrossRef]
10. Barbillon, G.; Ivanov, A.; Sarychev, A.K. Applications of Symmetry Breaking in Plasmonics. *Symmetry* **2020**, *12*, 896. [CrossRef]

11. Lim, D.-K.; Jeon, K.-S.; Kim, H.M.; Nam, J.-M.; Suh, Y.D. Nanogop-engineerable Raman-active nanodumbbells for single-molecule detection. *Nat. Mater.* **2010**, *9*, 60–67. [CrossRef] [PubMed]
12. Ding, S.-Y.; Yi, J.; Li, J.-F.; Ren, B.; Wu, D.-Y.; Panneerselvam, R.; Tian, Z.-Q. Nanostructure-based plasmon-enhanced Raman spectroscopy for surface analysis of materials. *Nat. Rev. Mater.* **2016**, *1*, 16021. [CrossRef]
13. Zambrana-Puyalto, X.; Ponzellini, P.; Maccaferri, N.; Tessarolo, E.; Pelizzo, M.G.; Zhang, W.; Barbillon, G.; Lu, G.; Garoli, D. A hybrid metal-dielectric zero mode waveguide for enhanced single molecule detection. *Chem. Commun.* **2019**, *55*, 9725–9728. [CrossRef] [PubMed]
14. Dolci, M.; Bryche, J.-F.; Leuvrey, C.; Zafeiratos, S.; Gree, S.; Begin-Colin, S.; Barbillon, G.; Pichon, B.P. Robust clicked assembly based on iron oxide nanoparticles for a new type of SPR biosensor. *J. Mater. Chem. C* **2018**, *6*, 9102–9110. [CrossRef]
15. Pichon, B.P.; Barbillon, G.; Marie, P.; Pauly, M.; Begin-Colin, S. Iron oxide magnetic nanoparticles used as probing agents to study the nanostructure of mixed self-assembled monolayers. *Nanoscale* **2011**, *3*, 4696–4705. [CrossRef]
16. Dolci, M.; Bryche, J.-F.; Moreau, J.; Leuvrey, C.; Begin-Colin, S.; Barbillon, G.; Pichon, B.P. Investigation of the structure of iron oxide nanoparticle assemblies in order to optimize the sensitivity of surface plasmon resonance-based sensors. *Appl. Surf. Sci.* **2020**, *527*, 146773. [CrossRef]
17. Barbillon, G.; Faure, A.C.; El Kork, N.; Moretti, P.; Roux, S.; Tillement, O.; Ou, M.; Descamps, A.; Perriat, P.; Vial, A.; et al. How nanoparticles encapsulating fluorophores allow a double detection of biomolecules by localized surface plasmon resonance and luminescence. *Nanotechnology* **2008**, *19*, 035705. [CrossRef]
18. Runowski, M.; Sobczak, S.; Marciniak, J.; Bukalska, I.; Lis, S.; Katrusiak, A. Gold nanorods as a high-pressure sensor of phase transitions and refractive-index gauge. *Nanoscale* **2019**, *11*, 8718–8726. [CrossRef]
19. Saboktakin, M.; Ye, X.; Chettiar, U.K.; Engheta, N.; Murray, C.B.; Kagan, C.R. Plasmonic Enhancement of Nanophosphor Upconversion Luminescence in Au Nanohole Arrays. *ACS Nano* **2013**, *7*, 7186–7192. [CrossRef]
20. Park, W.; Lu, D.; Ahn, S. Plasmon enhancement of luminescence upconversion. *Chem. Soc. Rev.* **2015**, *44*, 2940–2962. [CrossRef]
21. Wang, Y.; Ding, T. Optical tuning of plasmon-enhanced photoluminescence. *Nanoscale* **2019**, *11*, 10589–10594. [CrossRef] [PubMed]
22. Runowski, M.; Stopikowska, N.; Goderski, S.; Lis, S. Luminescent-plasmonic, lanthanide-doped core/shell nanomaterials modified with Au nanorods—Up-conversion luminescence tuning and morphology transformation after NIR laser irradiation. *J. Alloy. Compd.* **2018**, *762*, 621–630. [CrossRef]
23. Ding, S.-Y.; You, E.-M.; Tian, Z.-Q.; Moskovits, M. Electromagnetic theories of surface-enhanced Raman spectroscopy. *Chem. Soc. Rev.* **2017**, *46*, 4042–4076. [CrossRef] [PubMed]
24. Askes, S.H.C.; Schilder, N.J.; Zoethout, E.; Polman, A.; Garnett, E.C. Tunable plasmonic HfN nanoparticles and arrays. *Nanoscale* **2019**, *11*, 20252–20260. [CrossRef]
25. Manfrinato, V.R.; Camino, F.E.; Stein, A.; Zhang, L.H.; Lu, M.; Stach, E.A.; Black, C.T. Patterning Si at the 1 nm Length Scale with Aberration-Corrected Electron-Beam Lithography: Tuning of Plasmonic Properties by Design. *Adv. Funct. Mater.* **2019**, *29*, 1903429. [CrossRef]
26. Faure, A.C.; Barbillon, G.; Ou, M.; Ledoux, G.; Tillement, O.; Roux, S.; Fabregue, D.; Descamps, A.; Bijeon, J.-L.; Marquette, C.A.; et al. Core/shell nanoparticles for multiple biological detection with enhanced sensitivity and kinetics. *Nanotechnology* **2008**, *19*, 485103. [CrossRef]
27. Bryche, J.-F.; Gillibert, R.; Barbillon, G.; Sarkar, M.; Coutrot, A.-L.; Hamouda, F.; Aassime, A.; Moreau, J.; Lamy de la Chapelle, M. Density effect of gold nanodisks on the SERS intensity for a highly sensitive detection of chemical molecules. *J. Mater. Sci.* **2015**, *50*, 6601–6607. [CrossRef]
28. Dhawan, A.; Duval, A.; Nakkach, M.; Barbillon, G.; Moreau, J.; Canva, M.; Vo-Dinh, T. Deep UV nano-microstructuring of substrates for surface plasmon resonance imaging. *Nanotechnology* **2011**, *22*, 165301. [CrossRef]
29. Quilis, N.G.; Hageneder, S.; Fossati, S.; Auer, S.K.; Venugopalan, P.; Bozdogan, A.; Petri, C.; Moreno-Cencerrado, A.; Toca-Herrera, J.L.; Jonas, U.; et al. UV-Laser Interference Lithography for Local Functionalization of Plasmonic Nanostructures with Responsive Hydrogel. *J. Phys. Chem. C* **2020**, *124*, 3297–3305. [CrossRef]

30. Yang, L.T.; Lee, J.H.; Rathnam, C.; Hou, Y.N.; Choi, J.W.; Lee, K.B. Dual-Enhanced Raman Scattering-Based Characterization of Stem Cell Differentiation Using Graphene-Plasmonic Hybrid Nanoarray. *Nano Lett.* **2019**, *19*, 8138–8148. [CrossRef]
31. Bryche, J.-F.; Tsigara, A.; Bélier, B.; Lamy de la Chapelle, M.; Canva, M.; Bartenlian, B.; Barbillon, G. Surface enhanced Raman scattering improvement of gold triangular nanoprisms by a gold reflective underlayer for chemical sensing. *Sens. Actuators B* **2016**, *228*, 31–35. [CrossRef]
32. Barbillon, G.; Noblet, T.; Busson, B.; Tadjeddine, A.; Humbert, C. Localised detection of thiophenol with gold nanotriangles highly structured as honeycombs by nonlinear sum frequency generation spectroscopy. *J. Mater. Sci.* **2018**, *53*, 4554–4562. [CrossRef]
33. Chau, Y.F.C.; Chen, K.H.; Chiang, H.P.; Lim, C.M.; Huang, H.J.; Lai, C.H.; Kumara, N.T.R.N. Fabrication and Characterization of a Metallic-Dielectric Nanorod Array by Nanosphere Lithography for Plasmonic Sensing Applications. *Nanomaterials* **2019**, *9*, 1691. [CrossRef] [PubMed]
34. Farcau, C.; Marconi, D.; Colnita, A.; Brezestean, I.; Barbu-Tudoran, L. Gold Nanospot-Shell Arrays Fabricated by Nanoimprint Lithography as a Flexible Plasmonic Sensing Platform. *Nanomaterials* **2019**, *9*, 1519. [CrossRef] [PubMed]
35. Goetz, S.; Bauch, M.; Dimopoulos, T.; Trassi, S. Ultrathin sputter-deposited plasmonic silver nanostructures. *Nanoscale Adv.* **2020**, *2*, 869–877. [CrossRef]
36. Driencourt, L.; Federspiel, F.; Kazazis, D.; Tseng, L.T.; Frantz, R.; Ekinci, Y.; Ferrini, R.; Gallinet, B. Electrically Tunable Multicolored Filter Using Birefringent Plasmonic Resonators and Liquid Crystals. *ACS Photonics* **2020**, *7*, 444–453. [CrossRef]
37. Reguera, J.; Langer, J.; Jimenez de Aberasturi, D.; Liz-Marzan, L.M. Anistropic metal nanoparticles for surface enhanced Raman scattering. *Chem. Soc. Rev.* **2017**, *46*, 3866–3885. [CrossRef]
38. Barbillon, G.; Ivanov, A.; Sarychev, A.K. Hybrid Au/Si Disk-Shaped Nanoresonators on Gold Film for Amplified SERS Chemical Sensing. *Nanomaterials* **2019**, *9*, 1588. [CrossRef]
39. Langer, J.; Jimenez de Aberasturi, D.; Aizpurua, J.; Alvarez-Puebla R.A.; Auguié, B.; Baumberg, J.J.; Bazan, G.C.; Bell, S.E.; Boisen, A. Present and Future of Surface-Enhanced Raman Scattering. *ACS Nano* **2020**, *14*, 28–117. [CrossRef]
40. Li, Z.; Butun, S.; Aydin, K. Ultranarrow Band Absorbers Based on Surface Lattice Resonances in Nanostructured Metal Surfaces. *ACS Nano* **2014**, *8*, 8242–8248. [CrossRef]
41. Sarkar, M.; Besbes, M.; Moreau, J.; Bryche, J.-F.; Olivéro, A.; Barbillon, G.; Coutrot, A.-L.; Bartenlian, B.; Canva, M. Hybrid Plasmonic Mode by Resonant Coupling of Localized Plasmons to Propagating Plasmons in a Kretschmann Configuration. *ACS Photonics* **2015**, *2*, 237–245. [CrossRef]
42. Sobhani, A.; Manjavacas, A.; Cao, Y.; McClain, M.J.; Javier Garcia de Abajo, F.; Nordlander, P.; Halas, N.J. Pronounced Linewidth Narrowing of an Aluminum Nanoparticle Plasmon Resonance by Interaction with an Aluminum Metallic Film. *Nano Lett.* **2015**, *15*, 6946–6951. [CrossRef] [PubMed]
43. Yue, W.; Wang, Z.; Whittaker, J.; Lopez-Royo, F.; Yang, Y.; Zayats, A.V. Amplification of surface-enhanced Raman scattering due to substrate-mediated localized surface plasmons in gold nanodimers. *J. Mater. Chem. C* **2017**, *5*, 4075–4084. [CrossRef]
44. Safar, W.; Lequeux, M.; Solard, J.; Fischer, A.P.A.; Félidj, N.; Gucciardi, P.G.; Edely, M.; Lamy de la Chapelle, M. Gold Nanocylinders on Gold Film as a Multi-Spectral SERS Substrate. *Nanomaterials* **2020**, *10*, 927. [CrossRef] [PubMed]
45. Runowski, M.; Martin, I.R.; Sigaev, V.N.; Savinkov, V.I.; Shakhgildyan, G.Y.; Lis, S. Luminescent-plasmonic core–shell microspheres, doped with Nd^{3+} and modified with gold nanoparticles, exhibiting whispering gallery modes and SERS activity. *J. Rare Earths* **2019**, *37*, 1152–1156. [CrossRef]
46. Sheena, T.S.; Devaraj, V.; Lee, J.-M.; Balaji, P.; Gnanasekar, P.; Oh, J.-W.; Akbarsha, M.A.; Jeganathan, K. Sensitive and label-free shell isolated Ag NPs@Si architecture based SERS active substrate: FDTD analysis and -*Situ* Cell. DNA Detection. *Appl. Surf. Sci* **2020**, *515*, 145955. [CrossRef]
47. Wu, J.; Du, Y.; Wang, C.; Bai, S.; Zhang, T.; Chen, T.; Hu, A. Reusable and long-life 3D Ag nanoparticles coated Si nanowire array as sensitive SERS substrate. *Appl. Surf. Sci* **2019**, *494*, 583–590. [CrossRef]
48. Bryche, J.-F.; Bélier, B.; Bartenlian, B.; Barbillon, G. Low-cost SERS substrates composed of hybrid nanoskittles for a highly sensitive sensing of chemical molecules. *Sens. Actuators B* **2017**, *239*, 795–799. [CrossRef]
49. Magno, G.; Bélier, B.; Barbillon, G. Gold thickness impact on the enhancement of SERS detection in low-cost Au/Si nanosensors. *J. Mater. Sci.* **2017**, *52*, 13650–13656. [CrossRef]

50. Magno, G.; Bélier, B.; Barbillon, G. Al/Si nanopillars as very sensitive SERS substrates. *Materials* **2018**, *11*, 1534. [CrossRef]
51. Sarychev, A.K.; Ivanov, A.; Lagarkov, A.; Barbillon, G. Light Concentration by Metal-Dielectric Micro-Resonators for SERS Sensing. *Materials* **2019**, *12*, 103. [CrossRef]
52. He, Y.; Su, S.; Xu, T.T.; Zhong, Y.L.; Zapien, J.A.; Li, J.; Fan, C.H.; Lee, S.T. Silicon nanowires-based highly-efficient SERS-active platform for ultrasensitive DNA detection. *Nano Today* **2011**, *6*, 122–130. [CrossRef]
53. Barbillon, G.; Sandana, V.E.; Humbert, C.; Bélier, B.; Rogers, D.J.; Teherani, F.H.; Bove, P.; McClintock, R.; Razeghi, M. Study of Au coated ZnO nanoarrays for surface enhanced Raman scattering chemical sensing. *J. Mater. Chem. C* **2017**, *5*, 3528–3535. [CrossRef]
54. Lai, Y.C.; Ho, H.C.; Shih, B.W.; Tsai, F.Y.; Hsueh, C.H. High performance and reusable SERS substrates using Ag/ZnO heterostructure on periodic silicon nanotube substrate. *Appl. Surf. Sci* **2018**, *439*, 852–858. [CrossRef]
55. Yang, M.S.; Yu, J.; Lei, F.C.; Zhou, H.; Wei, Y.S.; Man, B.Y.; Zhang, C.; Li, C.H.; Ren, J.F.; Yuan, X.B. Synthesis of low-cost 3D-porous ZnO/Ag SERS-active substrate with ultrasensitive and repeatable detectability. *Sens. Actuators B* **2018**, *256*, 268–275. [CrossRef]
56. Lee, Y.; Lee, J.; Lee, T.K.; Park, J.; Ha, M.; Kwak, S.K.; Ko, H. Particle-on-Film Gap Plasmons on Antireflective ZnO Nanocone Arrays for Molecular-Level Surface-Enhanced Raman Scattering Sensors. *ACS Appl. Mater. Interfaces* **2015**, *7*, 26421–26429. [CrossRef]
57. Liu, X.W.; Wang, D.S.; Li, Y.D. Synthesis and catalytic properties of bimetallic nanomaterials with various architectures. *Nano Today* **2012**, *7*, 448–466. [CrossRef]
58. Song, J.B.; Duan, B.; Wang, C.X.; Zhou, J.J.; Pu, L.; Fang, Z.; Wang, P.; Lim, T.T.; Duan, D.W. SERS-Encoded Nanogapped Plasmonic Nanoparticles: Growth of Metallic Nanoshell by Templating Redox-Active Polymer Brushes. *J. Am. Chem. Soc.* **2014**, *136*, 6838–6841. [CrossRef]
59. Feng, J.J.; Wu, X.L.; Ma, W.; Kuang, H.; Xu, L.G.; Xu, C.L. A SERS active bimetallic core-satellite nanostructure for the ultrasensitive detection of Mucin-1. *Chem. Commun.* **2015**, *51*, 14761–14763. [CrossRef]
60. Zhang, Y.; Yang, P.; Habeeb Muhammed, M.A.; Alsaiari, S.K.; Moosa, B.; Almalik, A.; Kumar, A.; Ringe, E.; Kashab, N.M. Tunable and Linker Free Nanogaps in Core-Shell Plasmonic Nanorods for Selective and Quantitative Detection of Circulating Tumor Cells by SERS. *ACS Appl. Mater. Interfaces* **2017**, *9*, 37597–37605. [CrossRef]
61. Yang, Y.; Liu, J.; Fu, Z.W.; Qin, D. Galvanic replacement-free deposition of Au on Ag for core-shell nanocubes with enhanced chemical stability and SERS activity. *J. Am. Chem. Soc.* **2014**, *136*, 8153–8156. [CrossRef]
62. Matricardi, C.; Hanske, C.; Garcia-Pomar, J.L.; Langer, J.; Mihi, A.; Liz-Marzan, L.M. Gold Nanoparticle Plasmonic Superlattices as Surface-Enhanced Raman Spectroscopy Substrates. *ACS Nano* **2018**, *12*, 8531–8539. [CrossRef]
63. Di Martino, G.; Turek, V. A.; Tserkezis, C.; Lombardi, A.; Kuhn, A.; Baumberg, J. J. Plasmonic Response and SERS Modulation in Electrochemical Applied Potentials. *Faraday Discuss.* **2017**, *205*, 537–545. [CrossRef]
64. Luo, S.-C.; Sivashanmugan, K.; Liao, J.-D.; Yao, C.-K.; Peng, H.-C. Nanofabricated SERS-Active Substrates for Single-Molecule to Virus Detection in Vitro: A Review. *Biosens. Bioelectron.* **2014**, *61*, 232–240. [CrossRef]
65. Mosier-Boss, P.A. Review of SERS Substrates for Chemical Sensing. *Nanomaterials* **2017**, *7*, 142. [CrossRef]
66. Volk, K.; Fitzgerald, J.P.S.; Ruckdeschel, P.; Retsch, M.; König, T.A.F.; Karg, M. Reversible Tuning of Visible Wavelength Surface Lattice Resonances in Self-Assembled Hybrid Monolayers. *Adv. Opt. Mater.* **2017**, *5*, 1600971. [CrossRef]
67. Greybush, N.J.; Liberal, I.; Malassis, L.; Kikkawa, J.M.; Engheta, N.; Murray, C.B.; Kagan, C.R. Plasmon Resonances in Self- Assembled Two-Dimensional Au Nanocrystal Metamolecules. *ACS Nano* **2017**, *11*, 2917–2927. [CrossRef]
68. Cong, S.; Yuan, Y.; Chen, Z.; Hou, J.; Yang, M.; Su, Y.; Zhang, Y.; Li, L.; Li, Q.; Geng, F.; et al. Noble metal-comparable SERS enhancement from semiconducting metal oxides by making oxygen vacancies. *Nat. Commun.* **2015**, *6*, 7800. [CrossRef]
69. Liu, W.; Bai, H.; Li, X.; Li, W.; Zhai, J.; Li, J.; Xi, G. Improved Surface-Enhanced Raman Spectroscopy Sensitivity on Metallic Tungsten Oxide by the Synergistic Effect of Surface Plasmon Resonance Coupling and Charge Transfer. *J. Phys. Chem. Lett.* **2018**, *9*, 4096–4100. [CrossRef]

70. Lao, Z.; Zheng, Y.; Dai, Y.; Hu, Y.; Ni, J.; Ji, S.; Cai, Z.; Smith, Z.J.; Li, J.; Zhang, L.; et al. Nanogap Plasmonic Structures Fabricated by Switchable Capilarly-Force Driven Self-Assembly for Localized Sensing of Anticancer Medicines with Microfluidic SERS. *Adv. Funct. Mater.* **2020**, *30*, 1909467. [CrossRef]
71. Ghosh, P.; Paria, D.; Balasubramanian, K.; Ghosh, A.; Narayanan, R.; Raghavan, S. Directed Microwave-Assisted Self-Assembly of Au-Graphene-Au Plasmonic Dimers for SERS Applications. *Adv. Mater. Interfaces* **2019**, *6*, 1900629. [CrossRef]
72. Kuttner, C.; Höller, R.P.M.; Quintanilla, M.; Schnepf, M.J.; Dulle, M.; Fery, A.; Liz-Marzan, L.M. SERS and plasmonic heating efficiency from anisotropic core/satellite superstructures. *Nanoscale* **2019**, *11*, 17655–17663. [CrossRef] [PubMed]
73. Fusco, Z.; Bo, R.; Wang, Y.; Motta, N.; Chen, H.; Tricoli, A. Self-assembly of Au nano-islands with tuneable organized disorder for highly sensitive SERS. *J. Mater. Chem. C* **2019**, *7*, 6308–6316. [CrossRef]
74. Wu, P.; Zhong, L.-B.; Liu, Q.; Zhou, X.; Zheng, Y.-M. Polymer induced one-step interfacial self-assembly method for the fabrication of flexible, robust and free-standing SERS substrates for rapid on-site detection of pesticide residues. *Nanoscale* **2019**, *11*, 12829–12836. [CrossRef] [PubMed]
75. Yin, Z.; Zhou, Y.; Cui, P.; Liao, J.; Rafailovich, M.H.; Sun, W. Fabrication of ordered bi-metallic array with superstructure of gold micro-rings via templated-self-assembly procedure and its SERS applications. *Chem. Commun* **2020**, *56*, 4808–4811. [CrossRef] [PubMed]
76. Su, Y.; Xu, S.; Zhang, J.; Chen, X.; Jiang, L.-P.; Zheng, T.; Zhu, J.-J. Plasmon Near-Field Coupling of Bimetallic Nanostars and a Hierarchical Bimetallic SERS "Hot Field": Toward Ultrasensitive Simultaneous Detection of Multiple Cardiorenal Syndrome Biomarkers. *Anal. Chem.* **2019**, *91*, 864–872. [CrossRef]
77. Ning, C.-F.; Tian, Y.-F.; Zhou, W.; Yin, B.-C.; Ye, B.-C. Ultrasensitive SERS detection of specific oligonucleotides based on Au@AgAg bimetallic nanorods. *Analyst* **2019**, *144*, 2929–2935. [CrossRef]
78. Hussain, A.; Sun, D.-W.; Pu, H. Bimetallic core shelled nanoparticles (Au@AgNPs) for rapid detection of thiram and dicyandiamide contaminants in liquid milk using SERS. *Food Chem.* **2020**, *317*, 126429. [CrossRef]
79. Tian, S.; You, W.; Shen, Y.; Gu, X.; Ge, M.; Ahmadi, S.; Ahmad, S.; Kraatz, H.-B. Facile synthesis of silver-rich Au/Ag bimetallic nanoparticles with highly active SERS properties. *New J. Chem.* **2019**, *43*, 14772–14780. [CrossRef]
80. William, R.V.; Das, G.M.; Dantham, V.R.; Laha, R. Enhancement of Single Molecule Raman Scattering using Sprouted Potato Shaped Bimetallic Nanoparticles. *Sci. Rep.* **2019**, *9*, 10771. [CrossRef]
81. Joseph, D.; Kwak, C.H.; Huh, Y.S.; Han, Y.-K. Synthesis of AuAg@Ag core@shell hollow cubic nanostructures as SERS substrates for attomolar chemical sensing. *Sens. Actuators B* **2019**, *281*, 471–477. [CrossRef]
82. Vu, T.D.; Duy, P.K.; Chung, H. Nickel foam-caged Ag-Au bimetallic nanostructure as a highly rugged and durable SERS substrate. *Sens. Actuators B* **2019**, *282*, 535–540. [CrossRef]
83. Yilmaz, A.; Yilmaz, M. Bimetallic Core–Shell Nanoparticles of Gold and Silver via Bioinspired Polydopamine Layer as Surface-Enhanced Raman Spectroscopy (SERS) Platform. *Nanomaterials* **2020**, *10*, 688. [CrossRef] [PubMed]
84. Cai, J.; Li, Y.; Liu, C.; Wang, X. Green and Controllable Synthesis of Au–Ag Bimetal Nanoparticles by Xylan for Surface-Enhanced Raman Scattering. *ACS Sustain. Chem. Eng.* **2019**, *7*, 15154–15162. [CrossRef]
85. Prakash, O.; Sil, S.; Verma, T.; Umapathy, S. Direct Detection of Bacteria using positively charged Ag/Au Bimetallic Nanoparticles: A Label-Free Surface-Enhanced Raman Scattering Study Coupled with Multivariate Analysis. *J. Phys. Chem. C* **2020**, *124*, 861–869. [CrossRef]
86. Furlaz, A.; Almohammed, S.; Rice, J.H. Oxygen Incorporation-Induced SERS Enhancement in Silver Nanoparticle-Decorated ZnO Nanowires. *ACS Appl. Nano Mater.* **2020**, *3*, 1666–1673.
87. Yao, J.; Quan, Y.; Gao, M.; Gao, R.; Chen, L.; Liu, Y.; Lang, J.; Shen, H.; Zhang, Y.; Yang, L.; et al. AgNPs decorated Mg-doped ZnO heterostructure with dramatic SERS activity for trace detection of food contaminants. *J. Mater. Chem. C* **2019**, *7*, 8199–8208. [CrossRef]
88. Rajkumar, P.; Sarma, B.K. Ag/ZnO heterostructure fabricated on AZO platform for SERS based sensitive detection of biomimetic hydroxyapatite. *Appl. Surf. Sci.* **2020**, *509*, 144798. [CrossRef]
89. Zhu, Q.; Xu, C.; Wang, D.; Liu, B.; Qin, F.; Zhu, Z.; Liu, Y.; Zhao, X.; Shi, Z. Femtomolar response of a plasmon-coupled ZnO/graphene/silver hybrid whispering-gallery-mode microcavity for SERS sensing. *J. Mater. Chem. C* **2019**, *7*, 2710–2716. [CrossRef]

90. Wang, J.; Hassan, M.M.; Ahmad, W.; Jiao, T.; Xu, Y.; Li, H.; Ouyang, Q.; Guo, Z.; Chen, Q. A highly structured hollow ZnO@Ag nanosphere SERS substrate for sensing traces of nitrate and nitrite species in pickled food. *Sens. Actuators B* **2019**, *285*, 302–309. [CrossRef]
91. Liu, C.; Xu, X.; Wang, C.; Qiu, G.; Ye, W.; Li, Y.; Wang, D. ZnO/Ag nanorods as a prominent SERS substrate contributed by synergistic charge transfer effect for simultaneous detection of oral antidiabetic drugs pioglitazone and phenformin. *Sens. Actuators B* **2020**, *307*, 127634. [CrossRef]
92. Pal, A.K.; Pagal, S.; Prashanth, K.; Chandra, G.K.; Umapathy, S.; Mohan, D.B. Ag/ZnO/Au 3D hybrid structured reusable SERS substrate as highly sensitive platform for DNA detection. *Sens. Actuators B* **2019**, *279*, 157–169. [CrossRef]
93. Zhou, J.; Zhang, J.; Yang, H.; Wang, Z.; Shi, J.; Zhou, W.; Jiang, N.; Xian, G.; Qi, Q.; Weng, Y.; et al. Plasmon-induced hot electron transfer in Au–ZnO heterogeneous nanorods for enhanced SERS. *Nanoscale* **2019**, *11*, 11782–11788. [CrossRef] [PubMed]
94. Doan, Q.K.; Nguyen, M.H.; Sai, C.D.; Pham, V.T.; Mai, H.H.; Pham, N.H.; Bach, T.C.; Nguyen, V.T.; Nguyen, T.T.; Ho, K.H.; et al. Enhanced optical properties of ZnO nanorods decorated with gold nanoparticles for self cleaning surface enhanced Raman applications. *Appl. Surf. Sci.* **2020**, *505*, 144593. [CrossRef]
95. Graniel, O.; Iatsunskyi, I.; Coy, E.; Humbert, C.; Barbillon, G.; Michel, T.; Maurin, D.; Balme, S.; Miele, P.; Bechelany, M. Au-covered hollow urchin-like ZnO nanostructures for surface-enhanced Raman scattering sensing. *J. Mater. Chem. C* **2019**, *7*, 15066–15073. [CrossRef]
96. He, R.; Lai, H.; Wang, S.; Chen, T.; Xie, F.; Chen, Q.; Liu, P.; Chen, J.; Xie, W. Few-layered vdW MoO_3 for sensitive, uniform and stable SERS applications. *Appl. Surf. Sci.* **2020**, *507*, 145116. [CrossRef]
97. Hou, X.; Fan, X.; Wei, P.; Qiu, T. Planar transition metal oxides SERS chips: A general strategy. *J. Mater. Chem. C* **2019**, *7*, 11134–11141. [CrossRef]
98. Wang, X.; Li, J.; Shen, Y.; Xie, A. An assembled ordered $W_{18}O_{49}$ nanowire film with high SERS sensitivity and stability for the detection of RB. *Appl. Surf. Sci.* **2020**, *504*, 144073. [CrossRef]
99. Zhou, C.; Sun, L.; Zhang, F.; Gu, C.; Zeng, S.; Jiang, T.; Shen, X.; Ang, D.S.; Zhou, J. Electrical Tuning of the SERS Enhancement by Precise Defect Density Control. *ACS Appl. Mater. Interfaces* **2019**, *11*, 34091–34099. [CrossRef]
100. Wei, W.; Yao, Y.; Zhao, Q.; Xu, Z.; Wang, Q.; Zhang, Z.; Gao, Y. Oxygen defect-induced localized surface plasmon resonance at the WO_{3-x} quantum dot/silver nanowire interface: SERS and photocatalysis. *Nanoscale* **2019**, *11*, 5535–5547. [CrossRef]
101. Del Tedesco, A.; Piotto, V.; Sponchia, G.; Hossain, K.; Litti, L.; Peddis, D.; Scarso, A.; Meneghetti, M.; Benedetti, A.; Riello, P. Zirconia-Based Magnetoplasmonic Nanocomposites: A New Nanotool for Magnetic-Guided Separations with SERS Identification. *ACS Appl. Nano Mater.* **2020**, *3*, 1232–1241. [CrossRef]

 © 2020 by the author. Licensee MDPI, Basel, Switzerland. This article is an open access article distributed under the terms and conditions of the Creative Commons Attribution (CC BY) license (http://creativecommons.org/licenses/by/4.0/).

MDPI
St. Alban-Anlage 66
4052 Basel
Switzerland
Tel. +41 61 683 77 34
Fax +41 61 302 89 18
www.mdpi.com

Nanomaterials Editorial Office
E-mail: nanomaterials@mdpi.com
www.mdpi.com/journal/nanomaterials

www.ingramcontent.com/pod-product-compliance
Lightning Source LLC
LaVergne TN
LVHW070543100526
838202LV00012B/360